Water Quality Modelling for Rivers and Streams

Water Science and Technology Library

VOLUME 70

For further volumes:
http://www.springer.com/series/6689

To Anna and Mary

Preface

The idea of writing this book originated several years ago; however, due to certain factors it got delayed, and we have now finally been able to realise it. The objective of this book is to explain the computational methods that deal with the complex problem of water quality modelling in rivers and streams. The book provides in-depth coverage of computational open channel hydraulics as well as water quality issues and describes scientifically sound synthesised procedures that are relatively simple to use and fundamental for simulation purposes so that practical results can be achieved.

It was not our intention to replace any of the comprehensive books on numerical open channel hydraulics and water quality. The knowledge gathered in these two scientific areas over the last 50 years is so vast that it is impossible to review it comprehensively in a volume of this type.

On the contrary, the objective of the book is to address some fundamental problems of water quality in rivers and streams by integrating methods and procedures from these scientific fields. Our intention is thus to describe modelling fundamentals that will assist potential developers and users in devising the new generation of models, which will serve specific requirements in keeping with new legislation in many areas of the world, the Water Framework Directive in Europe being the principal reference.

Readers who would like to acquire deeper knowledge of the methods presented briefly in this book are recommended to refer to the following scientific literature on computational hydraulics: *Practical Aspects of Computational River Hydraulics*, by J. Cunge, F. Holly and A. Verwey; *Computational Hydraulics – Elements of the Theory of Free Surface Flow*, by M.B. Abbott; *Computational Techniques for Fluid Dynamics*, by C.A. Fletcher; and the recently published *Numerical Modelling in Open Channel Hydraulics*, by R. Szymkiewicz.

Readers can also refer to the following books on water quality science: *Aquatic Chemistry*, by W. Stumm and J. Morgan; *Surface Water-Quality Modeling*, by S.C. Chapra; *Hydrodynamics and Water Quality: Modelling Rivers, Lakes and Estuaries*, by J. Zhen-Gang; and *Quality Assurance for Water Analysis*, by P. Quevauviller.

Since this book is didactic, readers can consult even the more fundamental texts in river hydraulics, such as *Open Channel Hydraulics*, by V.T. Chow, and *Open Channel Flow*, by F.M. Henderson, which were the first introductory books on the subject published several years ago.

In view of the continuous advance in numerical techniques and computing facilities, there is an increasing production and circulation of ready-made software packages, which young engineers tend to use extensively. In light of this, it is also the aim of this book to explain the theoretical background for the simulation of the physical processes described, as well as the principles and limitations of these models, so that they can be applied effectively and safely to a variety of practical water quality problems.

It should be borne in mind that no model can offer useful results if it is not extensively calibrated and validated. Further, it should be remembered that successful model users are those who have the so-called engineering judgement.

The book is organised in 21 chapters and an appendix.

The first four chapters present introductory material related to the state of the art, the basics of pollution transport and the fundamental hydrodynamic processes.

Chapters 5, 6, 7, and 8 explain biochemical pollution and describe the most frequent pollutants in a river system.

Chapters 9, 10, 11, 12, 13, 14, and 15 are devoted to analytical and numerical methods, which can be used for simulating the pollution transport in rivers and streams.

Chapter 16 addresses thermal pollution and its simulation.

Chapters 17, 18, 19, and 20 are concerned with optimisation problems and the general processes of calibration and validation. They also deal with data acquisition and retrieval as well as with model reliability and measurement uncertainty.

Chapter 21 covers future trends and perspectives.

Last but not least, the appendix provides a short description of some widely used commercial ready-made packages.

We would like to express our gratitude to a number of people who assisted us in improving the content and the presentation of this book, in particular Prof. Mario Rosati of the University "La Sapienza" of Rome for his valuable suggestions and remarks in fundamental and applied mathematics. We would also like to thank Dr. D. Alexakis of the National Technical University of Athens for his assistance on quality issues and PhD candidate V. Bellos of the same university for running the numerical models with significant examples.

We would like to acknowledge the encouragement provided by members of the Permanent Working Group on "Water Quality" of the European Water Resources Association during the preparation of the book.

We would like to express our thanks to Ms. P. Van Steenbergen from Springer and to Prof. V.P. Singh, editor-in-chief of the series in which this book is included, for their suggestions during the finalisation of our work. We are also grateful to Ms. H. Vloemans, who helped us in preparing the text.

Finally, we would like to thank Prof. G. Viggiani of the University of Palermo and Prof. V.A. Tsihrintzis of the Democritus University of Thrace for their critical review of the manuscript and the encouraging appreciation of our work.

We hope that the reader will find the book practical, easy to follow and useful in professional life.

Rome, Italy   Marcello Benedini
Athens, Greece   George Tsakiris

About the Authors

Marcello Benedini graduated in civil engineering at the University of Padua, Italy, where he spent several years as an assistant to the chair of hydraulics, obtaining the official degree of professor. In 1969, he joined the Water Research Institute of the National Research Council and was appointed director of the water management sector. He carried out research on the advanced methods for the integrated use of water and environment protection, in collaboration with national authorities responsible for water management and with many other scientific institutions worldwide.

Benedini is active in Italian and international scientific organisations dedicated to water management problems.

He is also a co-founder of the European Water Resources Association (EWRA) and was recently recognised as honorary member.

He currently serves as editor-in-chief of the journal *European Water*.

George Tsakiris is a hydraulic engineer with a degree in Rural and Surveying Engineering from the National Technical University of Athens (NTUA), Greece, and a PhD in civil engineering from Southampton University, UK.

He is a professor in water resources management and engineering at the NTUA, director of the Laboratory of Reclamation Works and Water Resources Management and director of the Centre for the Assessment of Natural Hazards and Proactive Planning of NTUA.

He has convened many collaborative research projects in the area of water resources management as project coordinator and has published many research papers in international scientific journals and conferences.

He is currently president of the European Water Resources Association (EWRA) and editor-in-chief of the journal *Water Resources Management*.

Contents

1 Water Quality in the Context of Water Resources Management 1
  1.1 The Progress in Water Resources Management 1
  1.2 The Water Framework Directive 3
  1.3 EU Directives Related to Water Quality Issues 4
  1.4 From Pressures to Impacts 6
  References 9

2 Basic Notions 11
  2.1 Modelling and Water Quality Problems 11
  2.2 How to Interpret the Water Quality 12
  2.3 Water Resources Exploitation and Water Quality 14
  2.4 What to Model in the Water Quality Problems? 16
  2.5 The Most Common Way of Modelling 17
  2.6 General Features of the Mathematical Models 19
  2.7 Need of Data 20
  2.8 The River as the Main Water Body for Water
      Quality Protection 22
  2.9 A Fertile Field 24
  References 25

3 Mathematical Interpretation of Pollution Transport 27
  3.1 The Water Pollution 27
  3.2 The Advection Transport 29
  3.3 The Dispersion Transport: Fick Law 30
  3.4 External Contributions 32
  3.5 In-Water Transformations 32
  References 33

4 Fundamental Expressions 35
  4.1 The General Equation of Pollutant Transport in Water 35
  4.2 Nonconservative Pollutants 38
  4.3 Combined Processes 39
  4.4 Hydrodynamic Aspects 41
  4.5 A Simplified Interpretation 43
  References 47

5 Dispersion in Rivers and Streams 49
  5.1 The Importance of Dispersion 49
  5.2 Evaluation of the Dispersion Coefficient 50
  References 53

6 The Biochemical Pollution 57
  6.1 The Most Frequent Type of Pollution 57
  6.2 The BOD 58
  6.3 The Decay of BOD 60
  6.4 The Reaeration Coefficient 62
  6.5 The Saturation of Dissolved Oxygen 64
  6.6 Local Oxygenation Sources 64
  References 66

7 The Most Frequent Pollutants in a River 69
  7.1 Introduction 69
  7.2 The Oxygen Cycle 69
  7.3 Algae Activity 71
  7.4 Sediment Oxygen Demand 73
  7.5 Ammonia Oxidation 74
  7.6 Nitrite Nitrogen Oxidation 75
  7.7 The Nitrogen Cycle 75
       7.7.1 Organic Nitrogen 76
       7.7.2 Ammonia Nitrogen 76
       7.7.3 Nitrite Nitrogen 77
       7.7.4 Nitrate Nitrogen 78
  7.8 The Phosphorus Cycle 78
  7.9 Coliforms 80
  7.10 The Significant Constants 80
  7.11 Other Kinds of Pollutants 81
  7.12 Radioactive Substances 83
  7.13 Future Perspective and Research Needs 83
  References 84

8 Temperature Dependence 87
  8.1 Temperature Adjustment 87
  8.2 Heat Budget 88
  References 89

9 Application of the General Differential Equations 91
   9.1 An Analytical Solution 91
       9.1.1 Continuous Source of Infinite Duration 93
       9.1.2 Source of Finite Duration 95
   9.2 Some Comments 97
   9.3 Computing Procedures 98
   References 101

10 The Steady-State Case 103
   10.1 The Fundamental Equation in One-Dimensional Approach . . . 103
   10.2 The Nondispersive Flow 104
   References 107

11 Interpretation in Finite Terms 109
   11.1 Discrete Systems 109
       11.1.1 Advection 112
       11.1.2 Dispersion 112
       11.1.3 Nonconservative Pollutants 114
       11.1.4 Sinks and Sources 114
       11.1.5 The Demarcation Cross Sections of the Reaches 114
       11.1.6 The Fundamental Equation 115
       11.1.7 Combined Processes 116
   11.2 An Example 117
   11.3 Additional Comments 122
   References 124

12 Progress in Numerical Modelling: The Finite Difference Method . . 125
   12.1 Outlines of the Most Common Numerical Methods 125
   12.2 The Finite Difference Method (FDM) 126
   12.3 Basic Concepts of the Numerical Approach 127
   12.4 Numerical Schemes: Stability Criteria 129
   12.5 Boundary Conditions 131
   12.6 An Example: Pure Advection Transport 132
   12.7 Crank-Nicolson Numerical Scheme (CTCS) 135
   12.8 The BTCS Numerical Scheme 137
   12.9 Explicit Numerical Schemes 141
   12.10 Final Comments on the FDM 147
   References 148

13 The Finite Element Method 149
   13.1 Fundamental Aspects 149
   13.2 The Basic Algorithm for a Continuous Pollutant Injection 150
   13.3 The Ritz-Galerkin Approach 160
   13.4 An Application 167
   13.5 The Pollutant Wave 171
   13.6 Additional Comments on the Finite Element Method 176
   References 178

14   The Finite Volume Method   179
      14.1   Basic Concepts   179
      14.2   Additional Comments   182
      References   183

15   Multidimensional Approach   185
      15.1   The Two-Dimensional Case   185
      15.2   Examples   189
      15.3   An Outline of the 2D-Finite Element Method   193
      References   196

16   Thermal Pollution   199
      16.1   The Discharge of Hot Water   199
      16.2   The Basic Equations   200
      16.3   Point Heat Injection   202
      16.4   The Injection of a Hotter Flow   202
      16.5   Other Forms of Heat Injection   203
      16.6   Heat Exchange Between the River and Its Environment   204
             16.6.1   The Air-Water Heat Exchange   205
             16.6.2   Heat Exchange with the River Bed   206
             16.6.3   Heat Exchange with Sediments   207
      16.7   An Example   209
      References   211

17   Optimisation Models   213
      17.1   The Optimal River Management   213
      17.2   The Linear Programming Model   214
      17.3   Some Characteristics of the Linear Programming Models   215
      17.4   An Example of Linear Programming Model   216
      17.5   Post-optimal Analysis   218
      17.6   Other Programming Models   219
      17.7   The Role of Programming Models   219
      References   221

18   Model Calibration and Verification   223
      18.1   Calibration   223
      18.2   Verification   225
      18.3   Quantitative Model Performance Assessment   226
      References   228

19   Water Quality Measurements and Uncertainty   231
      19.1   Methods of Analysis   231
             19.1.1   Definite Methods   231
             19.1.2   Relative Methods   232
             19.1.3   Comparative Methods   232
             19.1.4   On-line Monitoring Methods   232

19.2   Traceability 233
19.3   Uncertainty 234
       19.3.1   Uncertainty of Sampling 234
       19.3.2   Sampling Variance Significant-Measurement
                 Variance Insignificant 234
       19.3.3   Sampling Variance Insignificant-Measurement
                 Variance Significant 236
       19.3.4   Sampling Variance Significant-Measurement
                 Variance Significant 236
       19.3.5   Uncertainty of Measurement 237
       19.3.6   Estimation of Total Uncertainty 238
19.4   Quality Assurance and Quality Control 239
19.5   Biological Indicators 240
19.6   Sediment and Suspended Solids 240
19.7   Physical Measurements 241
19.8   Hydrological Data 241
References 241

20 Model Reliability 245
20.1   The Effects of Parameter Variability 245
20.2   A More Refined Analysis 247
20.3   Uncertainty Analysis 251
20.4   The Monte Carlo Analysis 255
20.5   New Trends and Future Developments 261
References 262

21 Final Thoughts and Future Trends 265
21.1   New Developments 265
21.2   Future Trends 267
21.3   Epilogue 270
References 271

Appendix 273

Index 285

Chapter 1
Water Quality in the Context of Water Resources Management

Abstract Water resources management refers to all types of actions aiming at creating more favourable conditions for all water bodies in the future. This fundamental objective included in recent legislation, such as the Water Framework Directive of the European Union, brings water quality issues in the centre of interest of the water sector. Monitoring systems are now built in all European countries to produce original water quality data characterising the water bodies. Modelling techniques are complementary tools of great importance for assessing management decisions, which aim to improve the health of the water bodies. This book gives emphasis on innovative and classical methods useful for devising new comprehensive water quality models.

1.1 The Progress in Water Resources Management

Water is essential for life and for all human activities but also for preserving the environment and its resources. Rapidly growing population, intensification of agriculture, industrialisation, urbanisation, development of any kind and climatic factors are the main reasons for water scarcity conditions in many countries of the world.

The other side of the problem is the deterioration of water quality. Billions of people even today do not have access to safe water for drinking or other uses. The United Nations estimate that about 3,800 children die every day as a direct result of unsafe water and lack of sanitation.

Since gradually the available water is getting scarce, less food will be produced, more diseases will emerge and widespread poverty and hardship will prevail, sparking water conflicts in several regions of the world.

Water resources management is the scientific field that can assist in a rational equitable and efficient way of water resources development, treatment and use, safeguarding the sustainability of water resources and the environment.

M. Benedini and G. Tsakiris, *Water Quality Modelling for Rivers and Streams*,
Water Science and Technology Library 70, DOI 10.1007/978-94-007-5509-3_1,
© Springer Science+Business Media Dordrecht 2013

Sustainability, or *sustainable development*, has become household word since the report of the Brundtland Commission (World Commission on Environment and Development: Our common future, 1987), which states that the "Sustainable Development aims at ensuring that humanity meets its present needs without compromising the ability of future generations to meet their own needs". In short "sustainable development" implies that limitations should be imposed concerning the ability of the environment to fulfil the ever-increasing uses of resources so that "development" is able to last.

It is interesting to note that some decades ago countries and governments based their development on the approach of single-purpose planning. Each sector used its own criteria without considering the consequences of its decisions to the other sectors. Following this short-eyed approach, projects have been implemented having very negative consequences in other sectors, impeding the development. Even in the water sector, projects and measures have been decided creating catastrophic effects on other activities of the same sector.

As known, the definition of water resources management in the 1980s was referring to all activities aiming to fulfil the present and future water requirements with water of sufficient quantity and appropriate quality. This definition is lacking to secure the protection of the environment from any kind of abuses or natural hazards, since the main target is the fulfilment of demands. In other words, this led water resources management to become demand driven.

Focusing on water resources, it is evident that during the past decades some form of integration was attempted and adopted in development plans of most industrialised countries. The danger of destroying the environment led policy-makers and authorities to introduce (apart from the economic criteria) some rather vague environmental limitations of water abstractions and exploitation. Integration was also attempted in the process of management. The planning and construction phases were incorporated in the management process, whereas distinction was made between strategic and operational management.

In the late 1980s, the majority of parties concerned with the water sector (e.g. scientists, nongovernmental organisations, policymakers, river managers, authorities and all stakeholders) have adopted procedures presented as "integrated or comprehensive water resources management (IWRM)". Most laws and regulations of developed countries, in one form or the other, are influenced by the ideas of integration. As a logical consequence, water management systems became multi-objective using a series of criteria (e.g. economic, environmental, social). However, the most striking change in the new approach adopted in water resources management is the direct inclusion of water quality (or water pollution) in the management models (Quentin Grafton and Hussey 2011).

Methods now exist, incorporated in comprehensive management models, which can find the best possible development scenario by evaluating all accountable effects associated with it. Spatially, water resources management is applied to the entire river basin or watershed. According to Dzurik (1996), integrated water resources management is a specific application of the more general notion of integrated environmental management, which seeks to deal holistically with the natural environment.

Nowadays, it is widely understood that activities and processes in the watershed are linked close together in a continuum, which should be carefully modelled in order to assess any development scenario incorporating all the important activities related directly or indirectly to water resources and their quality. Perhaps the term water resources management does not clearly represent the new ideas that could be better represented by a term such as *Watershed Management*.

Points of water availability, centres of water consumption, lakes and rivers fragile environmental zones and ecosystems, sources of pollution and any other point of interest are linked together. The management aim is to achieve a favourable and stable relation between all these players.

In an even wider definition, the Watershed Management is replaced by the management of the water system, which comprises human, physical, biological and biochemical components (e.g. Craswell et al. 2007).

It should be stressed that sustainability is not another criterion in the multi-objective planning of the past. It is an extra requirement for the behaviour of the already existing criteria. Time series of criteria should be examined together with the additional limitation of being sustainable and covering all the above three components of the water system.

1.2   The Water Framework Directive

One of the most remarkable developments in the field of water resources management is no doubt the Water Framework Directive of the European Union, which has affected the legislation of many countries in the world (European Commission 2000).

The 2000/60 Directive establishes a framework for the protection of all waters (including surface waters, transitional and coastal waters and groundwaters).

This is achieved by:

1. Preventing further deterioration, protecting and improving the status of water resources
2. Promoting sustainable water use based on long-term protection of water resources
3. Protecting and improving the aquatic environments through reduction of discharges, emissions and losses of priority substances, and cessation or phasing out of discharges, emissions and losses of hazardous substances
4. Reducing the pollution of groundwater and preventing its further pollution
5. Mitigating the effects of floods and droughts

The key actions that the member states have taken for implementing the Water Framework Directive are:

(a) Identification of the river basins and formation of the river basin districts by spatial integration of adjacent river basins. Identification of responsible authorities (deadline 2003)
(b) Characterisation of river basin districts in terms of pressures, impacts and economics of water uses together with a register of protected areas (deadline 2004)
(c) Intercalibration of the ecological status classification systems (deadline 2006)
(d) Monitoring networks in operational mode (deadline 2006)
(e) Formulation of a programme of measures for achieving the environmental objectives of WFD in a cost-effective manner (deadline 2009)
(f) Presentation of River Basin Management Plans for each river basin district, including the designation of heavily modified water bodies (deadline 2009)
(g) Implementation of water pricing policies (deadline 2010)
(h) Rationalisation of the measures of the programme (deadline 2012)
(i) Implementation of programme of measures and achievement of environmental objectives (deadline 2015)

1.3   EU Directives Related to Water Quality Issues

As key factor in water resources management, the Water Framework Directive has favoured the implementation of measures so that the water quality in all water bodies is improved. Obviously the European Union has implemented a number of directives related to water quality and water pollution over the last decades. For informative reasons, the list of directives related to water quality issues follows. This information will assist any interested reader searching for specific data and criteria of water quality, applied in Europe and many other countries of the world:

Official journal of the European Communities, No. L 010: Directive of the European Parliament and of the council of 9 December 1996 on the control of major-accident hazards involving dangerous substances (96/82/EC).
Official journal of the European Communities, No. L 020: Council directive of 17 December 1979 on the protection of groundwater (80/68/EEC).
Official journal of the European Communities, No. L 031: Council directive of 16 June 1975 concerning the quality of bathing water (76/160/EEC).
Official journal of the European Communities No. L 103: Council directive of 2 April 1979 on the conservation of wild birds (79/409/EEC).

Official journal of the European Communities, No. L 123: Directive of the European Parliament and of the Council of 16 February 1998 on the placing of biocidal products on the market (98/8/EC).

Official journal of the European Communities, No. L 129: Council directive of 4 May 1976 concerning pollution caused by certain dangerous substances discharged in the aquatic environment of the Community (76/464/EEC).

Official journal of the European Communities, No. L 135: Council directive of 21 May 1991 concerning urban waste water treatment (91/271/EEC).

Official journal of the European Communities, No. L 162: Council directive of 21 May 1992 on the conservation of natural habitats and of wild fauna and flora (92/43/EEC).

Official journal of the European Communities No. L 182: Directive of the European Parliament and of the Council Ministers concerning waste landfills (99/31/EC).

Official journal of the European Communities, No. L 192: Commission decision of 17 July 2000 on the implementation of a European pollutant emission register (EPER) according to Article 15 of Council Directive 96/61/EC concerning integrated pollution prevention and control (IPPC) (2000/479/EC).

Official journal of the European Communities, No. L 194: Council directive of 16 June 1975 concerning the quality required of surface water intended for the abstraction of drinking water in the member states (75/440/EEC).

Official journal of the European Communities, No. L 222: Council directive of 18 June 1978 concerning the quality of fresh waters needing protection or improvement in order to support fish life (78/659/EEC).

Official journal of the European Communities, No. L 229: Council directive of 15 July 1980 relating to the quality of water intended for human consumption (80/778/EEC).

Official journal of the European Communities, No. L 230: Council directive of 15 July 1991 concerning the placing of plant protection products on the market (91/414/EEC).

Official journal of the European Communities, No. L 257: Directive of the European Parliament and of the council of 4 May 1976 concerning integrated pollution prevention and control (IPPC) (96/61/EC).

Official journal of the European Communities, No. L 281: Council directive of 30 October 1979 on the quality required of shellfish waters (79/923/EEC).

Official journal of the European Communities, No. L 327: Decision of the European Parliament and of the Council of 23 October 2000 establishing a framework for Community action in the field of water policy (2000/60/EC).

Official journal of the European Communities, No. L 330/32: Council directive 98/83/EC of 3 November 1998 on the quality of water intended for human consumption.

Official journal of the European Communities, No. L 375: Council directive of 12 December 1991 concerning the protection of waters against pollution caused by nitrates from agricultural sources (91/676/EEC).

1.4   From Pressures to Impacts

The WFD focuses on the pressures and impacts in Article 5, which requires for each river basin district:

(a) An analysis of its characteristics
(b) A review of the impacts of human activities on the status of surface waters and groundwater
(c) An economic analysis of water use

All the required tasks have been completed by many member states by 2004. Revision of those tasks is expected by 2013 and subsequently every 6 years. WFD initiated a process of assessment, iteration and refinement starting from the current conditions of each water body and forecasting the conditions at the end of the initial period (2015).

According to Annex II of WFD, the required review process is summarised in five tasks related to surface water bodies:

1. Characterisation of surface water body types
2. Definition of ecoregions and surface water body types
3. Establishment of reference conditions for each type of surface water body
4. Identification of pressures
5. Assessment of impacts

For a successful analysis of pressures and impacts, some prerequisites are required as shown in Fig. 1.1.

As anthropogenic pressures, WFD examines:

1. Point pollution sources
2. Diffuse pollution sources
3. Modification of flow regime (through abstraction or regulation)
4. Morphological alterations

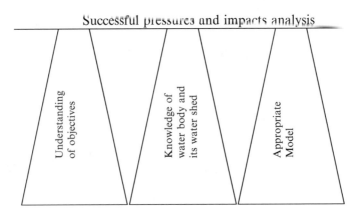

Fig. 1.1 The basic prerequisites for a successful development and application of appropriate mathematical models

In regard to the impact assessment, apart from the information derived from the pressures, additional information is needed (e.g. from environmental monitoring systems and/or necessary modelling procedures) to determine the most probable status of the surface water body in the future versus the set of environmental quality objectives and whether additional monitoring or a programme of measures is required.

All tasks related to pressures and impacts and the establishment of the resulting programme of measures (where it was required) had as the deadline the year 2009.

Although the impacts are the result of pressures, in many cases a systematic analytical framework is required for linking pressures and impacts. According to several researchers (e.g. CIS for WFD Guidance doc, no. 3) the DPSIR (driver, pressure, state, impact, response) analytical framework is already widely accepted and used.

An explanation of the DPSIR framework may be achieved through the explanation of terms included in the analysis:

Driver: activity which may have an environmental impact (e.g. industry)
Pressure: direct effect of the driver (e.g. change in water quality due to pollution from an industrial area)
State: physical, chemical and biological status of the water body
Impact: environmental effect of a pressure (e.g. fish death)
Response: measures taken to improve the state of the water body

An illustration of the DPSIR analytical framework appears in Fig. 1.2.

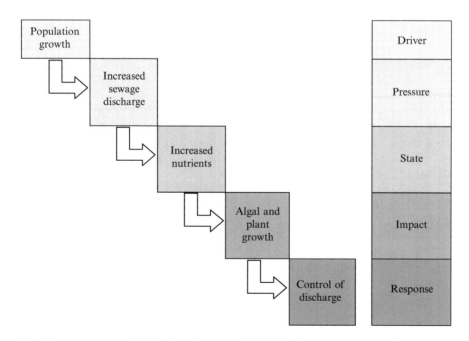

Fig. 1.2 An example of DPSIR (Modified from CIS WFD Guidance document no. 3)

The WFD defines three types of objectives for surface water bodies, namely, ecological status (for river, lakes transitional and coastal waters), ecological potential (for heavily modified or artificial water bodies) and chemical status (for all surface water bodies).

The categories of driving forces, pressures and impacts are presented in detail in CIS Guidance doc. 3. In the same document, a list of possible impacts or changes in state of surface water bodies that can be identified from monitoring data is presented as follows:

- Biological quality elements: macrophytes, phytoplankton, planktonic blooms, benthic invertebrates, fish, eutrophication
- Hydromorphological quality elements: hydrological regime, tidal, regime, river continuity, morphology
- Chemical and physicochemical quality elements: transparency, thermal conditions, oxygenation conditions, conductivity, salinity, nutrients status, acidification status, priority substances, other pollutants

The monitoring system at each water body can collect data on the quality items. However, no monitoring system can produce data of the appropriate density in space and time. To secure that data can be produced where they are needed, modelling techniques can be implemented as complementary tools to monitored data from the water body.

In particular, for the river system, numerous modelling attempts have been made since the original work of Streeter and Phelps in 1925. Although many of such river models suffer from the limitation that they use discrete point pollution sources instead of diffuse source inputs, they are still very important in contributing to the assessment of the water quality at each river segment. Although today there is a tendency for modelling not only the river system but the whole watershed, this book insists on giving emphasis on the processes and their modelling in the river system.

It can be easily understood that the new legislation on water resources management is very demanding and its application requires new advanced models to be devised and implemented in the river systems and the other water bodies.

As known, models provide the context in which decisions are made. Since the book is concerned with modelling, the limitations of the use of models should be always considered before they are used as decision support systems (Elms and Brown 2012a, b).

Examples of existing water quality models are included in the Appendix of this book. Unfortunately, existing models focus on classical pollution issues that can be computed and modelled. As explained, the main aim of this book is to analyse innovative and classical techniques and methods for devising future comprehensive water quality models for rivers and streams. According to some authors, rivers are among the most valuable but also the most abused resources on earth (Smits et al. 2000).

When reading this book, the basic knowledge on water quality issues is a prerequisite. Concerning the wide spectrum of topics related to water quality, the reader is advised to consult some of the numerous books on the subject (e.g. Stumm and Morgan 1996; Trimble et al. 2007; AWWA 1995; USEPA 2006).

References

AWWA (1995) Water quality, 3rd ed. Washington, DC

Craswell E, Bonell M, Bossio D, Demuth S, Van de Giesen N (eds) (2007) Integrated assessment of water resources and global change-a north-south analysis water resources management (special issue) 21(1):1–373

Dzurik A (1996) Water resources planning, 2nd edn. Rowman & Littlefield Publishes, Lanham, Colorado, USA: 345

Elms DG, Brown CB (2012a) Professional decisions: the central role of models. Civil Eng Environ Syst 29:165

Elms DG, Brown CB (2012b) Professional decisions: responsibilities. Civil Eng Environ Syst 29(3):176–190

European Commission (2000) Common implementation strategy for the water framework directive (2000/60/EC), Guidance documents

Quentin Grafton R, Hussey K (eds) (2011) Water resources planning and management. Cambridge University Press, Cambridge/New York, p 777

Smits AJM, Nienhuis PH, Leuven RSEW (eds) (2000) New approaches to river management. Backhuys Publishers, Leiden

Stumm W, Morgan JJ (1996) Aquatic chemistry, 3rd edn. Wiley, New York

Trimble SW, Stewart BA, Howell TA (2007) Encyclopedia of water science, 2nd edn. CRC Press, Boca Raton

USEPA (2006) Water quality standards review and revision. US EPA, Washington, DC

Chapter 2
Basic Notions

Abstract Mathematical models are now essential tools in water resources management and are currently applied for the solution of environmental problems including those of polluting discharge into surface and underground water bodies. Following the development of computational facilities and mathematical procedures, the models can provide reliable solutions, provided that they use proper data and are operated by competent professionals.

2.1 Modelling and Water Quality Problems

The progress of computer technology and mathematical procedures has introduced tools that are now essential for any activity of human life. This quite general and global aspect includes also the problems of water quality protection. The advantage of applying these tools for water quality control in rivers, lakes and aquifers is now appreciated by all scientists and professionals working in this field.

A tool that promises enormous power to address the water quality problems and give rational solutions is the mathematical model. According to Cox (2003), a water quality model *can mean anything from a single empirical relationship through a set of mass balance equations, to a complex software piece*. Much work has been done during the past decades, and today several mechanisms are available, not only in the scientific field but also in the flourishing software market. The use of mathematical models is now in the reach of any person who has a sufficient professional background to understand and deal with water quality problems. The mathematical models belong to a large family of models that is not new in the daily practice of water resources management and protection.

Because many aspects and variables have to be taken into account simultaneously dealing with water, the models are becoming more and more important in order to achieve reliable solutions in practical problems. The model acts as a representation of the reality and allows its problems to be handled without directly interfering with it. Verifying a solution directly in natural entities requires costly

M. Benedini and G. Tsakiris, *Water Quality Modelling for Rivers and Streams*,
Water Science and Technology Library 70, DOI 10.1007/978-94-007-5509-3_2,
© Springer Science+Business Media Dordrecht 2013

and complex engineering interventions, very often destructive. Vice versa, if the model represents correctly the reality and all its related phenomena, it can allow a solution to be examined in a short time and at a much lower cost.

As known, three types of models are currently used to solve water-related problems, namely, the hydraulic, the analogical and the mathematical models.

The *hydraulic* (*physical*) model consists of a reality at a different scale: a river stretch is reproduced by means of a duct having the same geometrical and morphological characteristics but in a size easy to be accommodated in the narrow space of the laboratory. With this model, the phenomena to analyse are the same as in reality (a flow in reality is reproduced by a flow in the model), and there is a plain correspondence between the various components (level, velocity, forces...), according to well-known *laws of similitude*.

The *analogical* models are based on a formal identity of the mathematical expressions that interpret different phenomena. Typical is the case of groundwater, for which the water flow is expressed by the Darcy law, formally identical to the Ohm law that interprets the electric current in a conducting line. After a suitable *scale of correspondence*, the behaviour of an aquifer, somewhat very difficult to analyse directly in the field, can be understood by the behaviour of an electric network having appropriate resistances and capacities.

The *mathematical models* interpret the reality by means of the numerical values that can be adopted to quantify the various phenomena and their components.

It is worthwhile to point out that the first two types of models just mentioned encounter now some drawback and are progressively abandoned in favour of the mathematical models, which are in fact more and more predominant. In the field of water quality, they are now probably the only effective tool.

Nevertheless, a lack of confidence still persists among several people, who think that the mathematical model is a too sophisticated mechanism, useful only for academic exercises but not in the real-world practice, where it has very often undergone unsuccessful outcomes.

To some extent, the use of mathematical model is a complex task, but it can assist in discovering the insight of a process if it is fed with reliable and proper data. The numerous successes during the last years have confirmed that when the mathematical model is in the hands of a skilled person with appropriate professional knowledge, it can give quite successful results. The mathematical model becomes then a device that helps the interested people to abide, step by step, with the ordinary way of thinking and to put into practice what they have learned with their daily experience.

2.2   How to Interpret the Water Quality

A typical incorrect use of water resources, which can cause dangerous effects to humans and other species, is the uncontrolled discharge of sewage into rivers and streams. It destroys the aquatic life and makes the water useless for any other use.

Table 2.1 The most significant quality indicators

1	– Temperature
2	– pH
3	– Dissolved oxygen (DO)
4	– Turbidity
5	– Conductivity
6	– Total organic carbon (TOC)
7	– Bacteria
8	– Viruses
9	– Chemical oxygen demand (COD)
10	– Biochemical oxygen demand (BOD)
11	– Metals and non-metals (Cr, Cd, Ni, As, Hg, Na, Br...)
12	– Phosphates
13	– Nitrogen compounds
14	– Organic compounds
15	– Suspended solids
16	– Salts (total dissolved salts)

Expensive treatments are then necessary, the cost of which becomes a burden for the whole community involved. Deterioration of river water has eventually an effect on the environment, on the human health and the economy.

The concept of water quality can be introduced by adopting some characteristic terms (generally called *parameters* or, better, *quality indicators*), which can be measured in the natural bodies and in the discharged water and are also characteristics of the water use. Table 2.1 lists some of the most important indicators, keeping in mind that such a list must be considered always open to introduce further terms that can be identified and detected by the research in progress. Each indicator can be measured, requires specific techniques of detection and analysis and imposes specific tasks for its control. Any use of water has its minimum and maximum values, determined after proper considerations.

As it will be better explained in the following chapters, the water quality in a river or stream depends on the quantity of water in which the pollutants are contained. It is, therefore, necessary that any action related to water quality is accompanied by accurate evaluations of the hydraulic conditions of the water body. In particular, water flow, level and velocity, which are determinant of pollutant behaviour, are to be carefully and frequently measured in the representative points of the water body. Moreover, the pollutants in water can be affected by rainwater and evaporation, and therefore, suitable measurements of the climatic and hydrological conditions are also of importance. Because the water quality in a natural or artificial body is a consequence of anthropogenic activities, the existing conditions of economic development, or the foreseeable trends, must be taken into consideration, with appropriate evaluation of all the terms to which a quality situation can be referred. Furthermore, water quality is controlled by natural factors that include geology and lithology of the watersheds and aquifers, the residence time, the reactions that take place within the aquifer and the type of land uses (Alexakis 2011).

2.3   Water Resources Exploitation and Water Quality

Among the several uses of water to be considered in the general framework of the "multipurpose" use of water resources, an important role is played by hydropower. It requires great quantities of water, but it is *not consumptive*, because the total amount withdrawn from a river is entirely returned to the natural bodies. Hydropower does not need special quality requisites, provided the natural turbidity is not so high as to cause siltation in the diversion ducts and erosion of pipes, valves and turbine blades.

Connected with the electricity generation is the use for thermal power generation. It is a nonconsumptive use, as the amount of water used for cooling and steam condensing is totally returned to the water body, even though with an increased temperature. To give an idea of the water quantity required for power generation, it is mentioned here that a thermoelectric plant of 660 MW demands about 20 m^3/s for a conventional generating group and about 28 m^3/s for a nuclear reactor. The increase of temperature, from the inlet to the outlet of the condenser, is about 4°C. Although the cooling water should be pure enough to avoid corrosion and crustation in the heat exchangers, particular adjustments in the materials used for the plant construction and an accurate running control allow also for water of poor quality to be used. Frequent is the case of using seawater, while a considerable quantity of freshwater can be saved for other uses demanding higher quality water. A small amount of water is also necessary to feed the boilers for steam generation. This water should be pure, and specific treatments are normally provided.

The agriculture requires huge quantities of water for irrigation. Only a small portion of the total amount delivered to the crops returns to the water bodies, while the largest portion is dispersed through evaporation and deep percolation in the soil, and a small quantity is transformed into the vital components of the plant. In agricultural activities, water can be used also for livestock and for farmer needs.

Irrigation does not require high-quality water, and in some cases, the required water can be produced using domestic and industrial wastewater, following old farming practices. Some salts can affect the growth of the crop and eventually accumulate in the plant with dangerous effects in case the plant or its fruits are used as food for animals and humans. Care must be taken for the use of wastewater containing bacteria, viruses or any kind of toxic substances. Some dissolved salts and suspended solids can alter the soil structure and modify the environmental conditions required by some vegetation species.

In an industrial factory, the water can be used for three different purposes:

– For *processing*, when it enters physically or chemically inside the composition of the final product. Normally, the quality must be high, achieved through appropriate treatment, the complexity and the cost of which depend on the original conditions in the water body and are characterised by the specific process adopted.

- For *cooling*, with more or less the same requirements pointed out for the thermal energy generation. The amount of such a use and the quality requirements are normally connected with the line of production and the size of the plants.
- For *washing*, in the final accomplishment of some production phases or for cleaning the plant premises. The amount and its quality depend on the size of the plant and are a function of the production process. Washing water is returned almost completely to the natural bodies with high pollution level.

The most important use of water is for *urban and domestic purposes*, for which it must have the highest degree of purity, especially from the sanitary viewpoint, and comply with the drinking water quality guidelines of several international organisations (WHO, EU, EPA, etc.). The amount to be supplied is evaluated in accordance with the degree of economical, social and technological development of the population to be served, ranging from a per capita 60 l/day in the case of small rural and scattered houses to a per capita 700 l/day for the largest urban communities.

Another use, although not well defined, is for *recreation*, which can vary in relation to the social and economic level of the population involved. Besides the quantity necessary for supplying vacation sites, hotels and holiday resorts (the quantity of which is determined following the rules of urban and domestic supply, keeping into account appropriate *peak* and *seasonal coefficients*), the recreation facilities are encountered in large water bodies suitable for boating, bathing and angling, with quality kept at the levels requested for the safe and effective exploitation of these uses.

Proper quantity and quality of water are necessary to maintain the original features of the landscape, historical places and sites of cultural heritage. An evaluation of the necessary amount of water for these uses is rather difficult and can be achieved following "ad hoc" investigations. The required level of water quality is normally high.

Due to the natural peculiarity of "washing out", streams and rivers are more and more frequently exploited to remove and dilute the waste discharged through the domestic and industrial sewers. Physical and biological processes contribute to the *self-purification process* so that the amount of pollutants is kept under proper thresholds. Specific values of the quality parameters are used in order to define the pollution level of the water bodies. If such values cannot be achieved through the natural dilution of the pollutants discharged, in order to secure possible further uses of water and to protect the aquatic life, the discharging water must undergo adequate treatment suitable to reduce the original pollutant concentration.

Water abstraction from a polluted natural body contributes to the modification of the original equilibrium of fauna and flora, motivating some irreversible alterations. Through the well-known *food chain*, life in the water environment can change, coming at the end to a threat for human survival. Protecting the water environment is, therefore, a necessity, especially for mankind's future, and has to be accomplished by maintaining the biological species originally existing in the water body biodiversity. Moreover, the protection of aquatic species allows sometimes the development of fisheries, which can be a prosperous activity.

Large rivers and canals are used for interior *navigation*, which assures low cost transportation especially regarding bulky raw materials. For this activity, river depth should be kept at proper levels and water velocity as low as possible, in order to allow for a safe ship motion and all other operations to be safely realised. While there are no requirements for water quality in the natural bodies, concentrated sources of pollution can be caused by accidental spills of fuel, lubricants or contaminating substances carried by the ship.

As seen in the preceding paragraphs, water can be used inside the river or stream (*in situ*) or after withdrawal (or *abstraction*) from the river and subsequent conveyance to another place. In the latter case, the original natural stream can be deprived of the amount of water that is necessary to maintain unaltered the original aquatic life and can prevent the possibility of other utilisations. To avoid such a risk, the regulations enforce in many countries require the fulfilment of a *minimum acceptable flow* in the river, which provides a threshold for limiting the water abstraction.

2.4   What to Model in the Water Quality Problems?

The main problem in water quality control and protection is learning how a pollutant is present in a water bulk and how its presence can vary in time and space. Such an evaluation is necessary in order to assess the attitude of a natural body for the environmental enhancement or in view of a feasible process useful to abate a dangerous contamination. The first step in dealing with a water quality problem is to provide reliable data relevant to the present pollutants. This falls under the responsibility of skilled people experts in geochemistry, analytical chemistry and biology. As already mentioned, a long list of pollutants is available, and the relevant analysis procedures are known, also in form of *standard methods* that can be officially recognised by institutions and authorities.

The reliability of water quality data depends not only on the precision of measuring instruments or the adopted analytical procedures but also on the way the samples are collected from the water body. Very often, a value is not representative of the river condition because it is relevant only to a particular aspect that does not involve the total bulk of the body. Repetition of measurements with direct statistical analysis and interpretation can provide a better data reliability. If a measurement cannot be repeated and only a single value can be collected, the data cannot be considered reliable. Data unreliability is generally due to several aspects, related to concurrent phenomena that cannot be always appreciated.

2.5   The Most Common Way of Modelling

For pollutants whose presence is appreciated in quantity (e.g. milligrams, micrograms), the *concentration*, expressed as mass per unit volume of water (e.g. milligram per litre), is the representative term. The pollutant concentration, C, is then the variable to be introduced in the model, as function of time and space:

$$C = C(x, y, z; t) \tag{2.1}$$

Considering a physical phenomenon, like the heat transfer in the water volume, the water temperature, τ, becomes the model variable:

$$\tau = \tau(x, y, z; t) \tag{2.2}$$

The water quality problem, to face by means of the model, is to find how the variable varies in the water body.

It is worthwhile to note that, once the problem is posed in mathematical form, the significant variable may lose its significance and become just an algorithm term, to be handled among many others, with the best computation procedure. In this context, there is a risk that the attention is brought principally to the mathematics, instead of the real problem. Such a risk has generated some misunderstanding, and very often the people who are competent in mathematics claim to be competent also in other disciplines (water pollution, air pollution and many problems of different nature, like the migration of fish in a river!), just because the significant variable dealt with in the model follows similar mathematical behaviour.

In water quality problems, the approach following the *Fick law* is the most widely adopted. It concerns, in particular, the process of dispersion (or diffusion), which is accompanied by other processes responsible for the migration of pollutants in the water volume, as it will be explained in Chap. 3.

Generally speaking, even some simple correlation of experimental data can be considered as mathematical models. This is so because the analyst can learn from the statistical manipulations and can interpret the real situation without entering into the inner mechanisms of pollutant transport. However, advanced technology and the new way of approaching and interpreting a water quality problem have made the mathematical model a device that can be applied in all the aspects encountered in practice.

Some well-known and widely used mathematical models for water quality in rivers and catchments are listed in Table 2.2, with the indication of the institutions in which they have been developed. These institutions can provide sufficient details about the model structure; in the table, there are also some references useful for the reader, for which they provide the general description, the principles and some applications of these models.

The scientific and technical literature is lavish in providing textbooks and notes on the most recent research findings of water quality models. It might be interesting to recall some original attempts of model development to have an idea of the basic approach and the progress achieved so far.

Table 2.2 Principal water quality models for rivers and streams

Country	Institution	Year	Model name	Purpose	References
USA	USCE	1982	CE-QUAL	Substance transport and transformation	Wells (2000)
Netherlands	DH	1985	DELWAQ	Pollution transport	Delft (1990) and Delft Hydraulics (1992)
USA	USEPA	1987	QUAL2E	Pollution transport	Brown (1987) and Brown and Barnwell (1987)
France-UK	LNH-CEH	1991	TELEMAC	Water flow and pollution transport	Kopman and Markofsky (2000) and Galland et al. (1991)
Switzerland	EAWAG	1994	AQUASIM	Substance transport and transformation	Reichert (1998)
UK	CEH	1997	PC-QUASAR	Water flow and pollution transport	Lewis et al. (1997) and Whitehead et al. (1997)
Denmark	DHI	1999	MIKE 11	Water quality and sediment transport	Hanley et al. (1998)
UK	Newcastle University	2008	TOPCAT-NP	Simulation of flow and nutrient transport	Quinn et al. (2008)
Germany	IGB	2009	MONERIS	Regionally differentiated quantification of nutrient emissions into a river system	Venohr et al. (2009)
UK	EA	2010	SIMCAT	Fate and transport of solutes	Warn (2010)

In this respect, fundamental is the work done by Thomann (1971) to understand and interpret the basic concepts of water quality problems. Interesting applications were described by Beck (1983a, b). Biswas (1981) collected some interesting contributions in a fundamental textbook.

Up to now, a fully comprehensive review has been produced by Chapra (1997), also with the support of a very long reference list. A first general set of considerations was presented originally by Raunch et al. (1998). Two recent reviews of some water quality models also present the recent developments in the field of popular quality modelling (Cox 2003; Tsakiris and Alexakis 2012).

Some of these models are reviewed in the Appendix of this book.

The water quality problems and the role of mathematical models are now one of the main subjects of the scientific research, and consequently, new remarkable contributions can be expected in the qualified journals, while refined advanced software packages are also expected by engineering firms.

2.6   General Features of the Mathematical Models

The recent development in computer science and mathematics has pointed out many particular aspects of the mathematical models that should be briefly recalled, in order to better understand the model structure and its working mode. Great contribution to this knowledge has been given by the *systems science*, which is now useful to interpret and solve the very complex problems of the modern era and for which the mathematical models are some of the most efficient tools.

The fundamental point in the description of a model is the identification of its constituent terms. These can be grouped into three categories, namely:

- *Constants*, which maintain always the same value for the entire application and are determined in a way that is independent on the inner mechanisms of the model
- *Variables*, the numerical value of which can change one or several times during the model application
- *Parameters*, the numerical value of which is arbitrarily fixed for some steps of the application

Some variables describe at every moment the evolution of the problem dealt with: they are the *state variables*. For some others, the numerical value can be altered, although always complying with the model rules: they are the *decision variables*. The initial value given to some variables makes up the *model input,* and consequently, these are the *input variables*; the value inferred from the other variables through the model application is the *model output*.

A primary role is played by the formulae or relationships connecting the variables, able to produce a certain output for a given input. Some formulae, able to clearly interpret the phenomena for which the model is developed, are available in the scientific and technical literature. Otherwise, the formulae can be determined through a statistical interpretation of the available data. Consequently, two large categories of mathematical models can be distinguished: in the first category, the relationship among the variables is a mathematical function, and only one output value (or set of values) corresponds to a given input value (or set of values). Such a model is called *deterministic*. In the second category, the relationship among the variables is based on the probability of occurrence, and the mathematical model becomes *probabilistic* (or *stochastic*). A distinction between probabilistic and stochastic is often pointed out, as the latter term is used generally when the time is one of the variables. Both deterministic and probabilistic models are currently used in the water problems, and in many cases, the same problem has to be treated in both ways.

The mathematical models that represent the reality in descriptive way are the *simulation models*. Their substantial feature is a "translation" in mathematical terms of the evolution phases of the reality.

There are also some models able to give, for a given input, an output that can be considered, within a certain respect, the best one among a set of possible values: they are the *optimisation* (or *programming*) *models,* and their substantial point lies on the identification of an *objective function* to be optimised with the involved variables and in view of a predetermined criterion of operation.

Generally, a problem, particularly in the water context, cannot be thoroughly examined by means of a single model, but several models are often necessary, both of simulation and optimisation, interconnected, in order to achieve a reliable result.

2.7   Need of Data

It is worthwhile to stress again that the mathematical models remain a useless tool if there are no suitable values for the terms involved in their application and that they can show their power only with the availability of proper data. The collection and the way of making data available play a very important role in model application. Any consideration about data implies the highest professional experience of the people working with models, who are responsible to collect them and evaluate their possibility to comply with the requirements of the problem to which the model is concerned.

Mathematical and statistical theories have produced techniques that, with the help of the advanced computing facilities, can be useful in treating the available data, presenting them in a form useful for their insertion into the mathematical models. Such techniques consist essentially of special computer languages or packages that allow the data:

– To be continuously adjourned with the results of the measurements becoming available in the meantime
– To be saved without disturbing the natural interaction existing among them
– To be retrieved according to the specific requests of the user

These techniques allow for the construction of the *data bases*. Getting data available for the application of mathematical models is the concern of some activities and tools that make up the *data banks*, words that reflect the well-known institutions where every client can deposit and cash his money. The data bank consists of the following steps:

• *Data collection*, which is based (1) on machines (*hardware*) able to transfer the values measured at the gauges into a form which can be accepted by the computers, (2) on the appropriate mathematical programmes (software) and (3) on specialised expertise (technicians, software engineers, etc.)
• *Data screening*, based principally on statistical evaluations (calculations of central values, extreme values, etc.)

Fig. 2.1 How data are necessary and can be used for a model

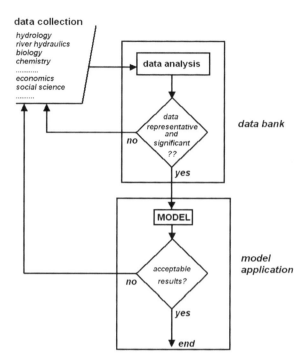

- *Data saving*, requiring the proper structure of data bases and the necessary computing facilities
- *Data retrieval*, in favour of the users, relying on appropriate hardware (printer, plotter, hard disk drive, etc.), with adequate software for several operations including presentation

In the problems of water resources management, the data banks should be run preferably by the same people who have the responsibility to intervene for water protection and utilisation. Access to data, however, should be given to all the interested institution and stakeholders.

The way data are considered for the model application is illustrated in Fig. 2.1. Once the problems are clearly defined, data must be collected in all the pertinent disciplinary sectors, with the best possible accuracy, using the proper instrumentation and trying to benefit from the most advanced professional experience. After developing the various steps for constructing the data bank described above, there is the need to verify whether the collected data can respond to the requests of the problem, in a significant and representative way. If the result is not satisfactory, new data must be collected, following criteria and procedures suggested by the need to have more significant details and to better interpret the problem. Refined and acceptable data can then be used for the model, but also at this step, there is a need for other investigations to ascertain that the data can be satisfactory for the proper model application. Otherwise, new and much more refined data are necessary.

The same figure underlines the necessity and the convenience of repeating steps, in order to obtain data that can be acceptable for phenomena interpretation and for an efficient use of the model.

2.8 The River as the Main Water Body for Water Quality Protection

Environmental protection concerns all the water bodies existing in nature, namely, rivers, lakes, lagoons, coastal water and groundwater; the quality of them should be in relation with more general aspects relevant to the living conditions of the involved people.

The rivers characterise the importance of all the natural water bodies, because they fill and empty lakes and lagoons, recharge groundwater and eventually discharge into the coastal water. Rivers play a primary role in assessing the availability of natural resources, supplying the water necessary to the various uses. Monitoring and controlling the river quality is, therefore, an essential and unavoidable step in water resources management.

Both in situ and abstraction water uses have to refer to the amount of water that the river is able to supply, in terms of quantity and quality, taking into account the protection of the environment. Maintaining an acceptable quality in the river can assure the correct resources exploitation, allowing for competitive uses to be fulfilled in the most rational way, according to the most up-to-date view of an integrated resources management.

The fundamental theory of water quality models described in the following chapters is valid for any type of water body. However, several adjustments can transform a water quality model developed for a river into models for stagnant water bodies or for aquifers. The models described in these pages refer to all kinds of bodies with running free surface water, like natural rivers, streams and artificial canals.

To perform a management activity in an effective manner, an appropriate definition of the water resources size and peculiarities is necessary. First of all, adequate space boundaries have to be set, and the most logical way to do this is to ascertain what sort of physical and non-physical ties exist among the various parts of which the resource is built up. Generally, the water belonging to a stream cannot be considered apart from the spring from which the stream originates or from rainwater travelling on the slopes that contributes to the river in question and apart from the water body into which it enters. All the physical aspects relating in one or other way with these quantities of water can find a unitary binding in the *river basin* (also *hydrographic basin*), defined as the whole of water bodies contributing to build up a unique river or, better, the area from which all the water molecules, fallen as natural rainfall or introduced from elsewhere both naturally and artificially, are brought by natural flow to a unique cross section, from which the river discharges into the sea or a great lake. Such a definition is often equivalent to that of *catchment area*, although the latter term is more

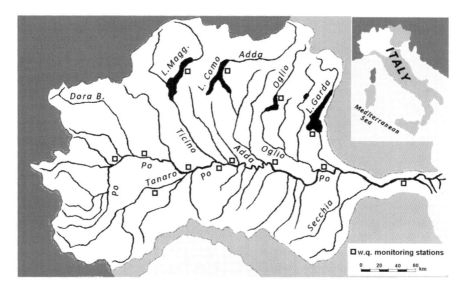

Fig. 2.2 Location of monitoring stations in a large river basin (River Po in Italy)

preferably applied to the area in which the run-off is made up only by the rainfall and other types of precipitation, while the term hydrographic basin, including also groundwater, is more comprehensive. However, as in most publications, the concepts of river basin or catchment are used interchangeably.

From the above, it is understood that it is difficult for a water quality model to study only the river segments and their conditions without studying the processes of water encountered in the entire river basin. This is the reason why most models refer to both the basin and the river.

The management of the water resources in a river basin is also facilitated because very often non-physical aspects are common in it and are distinguished from those that can be found in another basin. Mainly in the case of large rivers, the basin boundaries (*hydrographic divide*) enhance discrimination of life customs and trends of activities; sometimes, people dwellings in the same basin are bound by old and enduring traditions, even though they belong to different states or administrations.

These hydrologic and physical considerations support the identification of a *river authority*, responsible of all the aspects referring to the basin, including water quantity management, water quality control and environmental protection. According to the most current views of an integrated water resources management, the river authority is responsible of all the processes and episodes occurring to water-related aspects in the basin, providing infrastructure tools and financial resources to intervene in order to achieve the optimal rational use of the available water, to prevent inundation and damages caused by floods and to control the water quality and protect the environment. Concerning the water quality problems, the authority should provide suitable *monitoring systems*, collecting data in the most representative points of the basin. In large basins, as in the example shown in Fig. 2.2, the data collection can be a very demanding and costly activity.

2.9   A Fertile Field

During the last decades, the modelling practice has gained a remarkable acceptance in solving real-world problems, as confirmed by the numerous applications promoted by the responsible authorities. A remarkable progress is appreciable if the first attempts to review this area (Ray 1988; Van Pagee 1984) are compared with the more recent publications (Cox 2003; Tsakiris and Alexakis 2012). The technical staff has acquired confidence with the models and their intrinsic powerful characteristics, removing the original scepticism that had characterised the first attempts of application of models as were promoted by the scientific community. The responsible authorities have adopted the model as an essential tool able to support the decision making process, without being necessarily aware of the inner mechanism on which the model is developed and works. The model output is the main answer that is requested after an application carried out using a series of data representative of the application.

The technical literature reports successful examples, relevant to the largest river basins in the world, like that of the Po (Delft 1990), the Nile (Hafez 2003), and the Senne (Van Griensven and Bauwens 2002). The description of these applications underlines the importance of the basic information of the river water quality on which the model is constructed (Harmancioglu 1991). The role of the model in a more general context of water resources management is frequently underlined, also in relation to the design and construction of huge works, like the large dam on the Yellow River (Jinxiu et al. 2001; Yangwen et al. 2007), but also to well-defined goal to which the model is oriented, like the sanitation of the urban environment (Paoletti et al. 2004). Worthy to be mentioned, in particular, is the attempt to combine the pollution transport with the hydraulic aspects of the river, both in steady normal conditions (Schaffranek 1998) and in the case of flood events (Koussis 1983).

Remarkable progress can be observed also in the scientific interest, and a long path has been covered since the beginning. The comprehensive studies carried out just a few decades ago (Stanbury 1986) can be considered tiny in comparison with the more recent developments that will be described in the next pages, even though they can still be considered the reference point of the flourishing research that can be recorded at the present time. The above considerations justify the scientific interest for the subject of water quality models and can be a valid incentive to continue and go deeper and deeper in the study of this matter.

This interest is fostered by the need for producing water quality results required by the governments in order to achieve the goal of better quality water resources in the future, as highlighted in the first chapter of this book, in which the emphasis was given to the implementation of the WFD (Dir.2000/60) in the member states of the European Union.

References

Alexakis D (2011) Diagnosis of stream sediment quality and assessment of toxic element contamination sources in East Attica, Greece. Environ Earth Sci 63(6):1369–1383

Beck MB (1983a) A procedure for modeling. Mathematical modeling of water quality: streams, lakes, and reservoirs. In: Orlob G (ed) International institute for applied systems analysis, Laxenburg, Austria, pp 11–41

Beck MB (1983b) Application of water quality models. Mathematical modeling of water quality: streams, lakes, and reservoirs. In: Orlob G (ed) International institute for applied systems analysis, Laxenburg, Austria, pp 425–467

Biswas AK (ed) (1981) Models for water quality management, Mc Graw-Hill series in water resources and environmental engineering. Mc Graw-Hill, New York

Brown LC (1987) Uncertainty analysis in water quality modelling using QUAL2E. In: Systems analysis in water quality management. Pergamon Press, New York

Brown LC, Barnwell TO (1987) The enhanced stream water quality models QUAL2E and QUAL2E-UNCAS: documentation and user model, EPA-600/3-87/007, US EPA, Athens, GA

Chapra SC (1997) Surface water-quality modeling. WCB-McGraw Hill, Boston

Cox BA (2003) A review of currently available in-stream water quality models and their applicability for simulating dissolved oxygen in lowland rivers. Sci Total Environ 314–316:335–377

Delft Hydraulics (1990) Water quality analysis of the Po river. Report international T690. Delft, The Netherlands

Delft Hydraulics (1992) The water quality model DELWAQ. Release 3.0, User's manual, Deft, The Netherlands

Galland JC, Goutal N, Hervouet JM (1991) TELEMAC: a new numerical model for solving shallow water equations. Adv Water Resour 14(3):138–148

Hafez YI (2003) Some aspects of water quality modeling for the Nile river from Aswan to Cairo, Egypt. In: Proceedings of the 30th IAHR congress, Thessaloniki, Greece, theme B, pp 715–722

Hanley N, Faichney R, Munro A, Shortle JS (1998) Economic and environmental modelling for pollution control in an estuary. J Environ Manage 52(3):211–225

Harmancioglu NB (1991) An information-based approach to monitoring and evaluation of water quality data. In: Tsakiris G (ed) Proceedings of the European conference advances in water resources technology, Athens, Greece. Balkema Publisher, Rotterdam, The Netherlands, pp 377–386

Jinxiu L, Dehui S, Jinchi H, Wengen L, Xuezhong Y, Jing P (2001) Prediction of water quality for the three Gorges reservoir. In: Proceedings of the 29th IAHR congress, Beijing, China, theme B, pp 717–722

Kopman R, Markosfky M (2000) Three–dimensional water quality modelling with TELEMAC-3D. Hydrol Process 14(143):2279–2292

Koussis AD (1983) Unified theory for flood and pollution routing. J Hydraul Eng ASCE 109 (12):1652–1664

Lewis DR, Williams RJ, Whitehead PG (1997) Quality Simulation along Rivers (QUASAR): an application to the Yorkshire Ouse. Sci Total Environ 194–195:399–418

Paoletti A, Sanfilippo U, Innocenti I (2004) Advances in modelling the experimental data on pollutant dynamic in ephemeral streams influenced by transient storage in dead zones. In: Marsalek J et al (eds) Enhancing urban environment by environmental upgrading and restoration. Kluwer Academic Publishers, Dordrecht, pp 289–306

Quinn PF, Hewett CJM, Dayawansa NDK (2008) TOPCAT-NP: a minimum information requirement model for simulation of flow and nutrient transport from agricultural systems. Hydrol Process 22:2565–2580

Raunch W, Henze M, Koncsos L, Reichert P, Shanaham P, Somlyody L, Vanrolleghem P (1998) River water quality modelling: 1. State of the art. Proceedings of the IAWQ biennial conference, Vancouver, Canada

Ray C (1988) In: Biswas AK, Khoshoo TN, Khosla A (eds) Use of QUAL-I and QUAL-II models in evaluating waste loads to streams and rivers: environmental modelling for developing countries. Tycooly Publishing, London, pp 65–79

Reichert P (1998) AQUASIM – a tool for simulation and data analysis of aquatic systems. Water Sci Technol 30(2):21–30

Schaffranek RW (1998) Simulation model for open-channel flow and transport. In: Proceedings of the first federal interagency hydrologic modeling conference, Las Vegas, NV: subcommittee on hydrology of the Interagency Advisory Committee on water data, pp 1–103 to 1–110

Stanbury J (ed) (1986) Water quality modelling in the inland natural environment. Papers presented at the international conference, Bournemouth, England. British Hydromechanics Research Association, Cranfield

Thomann RV (1971) Systems analysis and water quality management. Environmental Science Service Division, New York

Tsakiris G, Alexakis D (2012) Water quality models: an overview. European Water 37:33–46

Van Pagee JA (1984) Water quality modelling of the river Rhine and its tributaries in relation to sanitation strategies. Water Sci Technol 16(3):393–406

Van Griensven A, Bauwens W (2002) River water quality management for the Senne river Basin (Belgium). In: Tsakiris G (ed) Proceedings of the 5th EWRA international conference water resources management in the era of transition, Athens, Greece, pp 520–528

Venohr M, Hirt U, Hofmann J, Opitz D, Gericke A, Wetzig A, Ortelbach K, Natho S, Neumann F, Hurdler J (2009) The model system MONERIS version 2.14.1vba manual. Leibniz Institute of Freshwater Ecology and Inland Fisheries in the Forschungsverbund Berlin e.V., Müggelseedamm 310, D-12587 Berlin, Germany

Warn T (2010) SIMCAT 11.5 a guide and reference for users. Environment Agency. 111_07_SD06 Rotherham, UK

Wells SA (2000) Hydrodynamic and water quality river basin modelling using CE-QUAL-WS version 3. In: Ibarra-Bastegi G, Brebbia C, Zannetti P (eds) Development and application of computer techniques to environmental studies. WOT Press, Boston, pp 195–204

Whitehead PG, Williams RJ, Lewis DR (1997) Quality simulation along River Systems (QUASAR): model theory and development. Sci Total Environ 194–195:447–456

Yangwen J, Cunwen N, Hao W, Hui G, Suhui S (2007) Integrated assessment of water quantity and quality in the Yellow River basin. In: Proceedings of the 32nd IAHR congress, Venezia, Italy, vol 1, p 310 (printed summary)

Chapter 3
Mathematical Interpretation of Pollution Transport

Abstract The pollutant fate and transport in a river or stream is governed by some physical and chemical processes, which can be formally interpreted considering an elementary volume of the water body. Some principal expressions of pollutant concentration are formulated, which will be useful for the development of the mathematical models. Evaluating the pollutant concentration in the water volume assists in finding how the contact of pollutant with water controls its final quality.

3.1 The Water Pollution

One of the main tasks in environmental protection is the control of water quality in relation to the pollution caused by the discharge of urban, agricultural and industrial wastes. The "pollutants" affecting the water quality in running and standing water bodies have physical, chemical and/or biological characteristics. These characteristics can be determined by means of analytical tools that are now very precise.

Some more detailed considerations on these aspects will be addressed in Chap. 7 and are contained in several specialised books (Foster et al. 1995; Gray 2008; Trudgill et al. 1999; Quevauviller 2002).

The presence of pollutants in water can be exhibited in various forms. It can be in form of *suspension*, in which the pollutant particles have solid status. It can be in form of *chemical solution*, a homogeneous mixture of solids, liquid or gaseous pollutant and water particles. It can be also in form of *emulsion*, if the pollutant is *immiscible* with water. In all cases the considerations developed in these chapters will not make any distinction, because the mentioned forms of presence concern phenomena at a different scale. It is only worthy to mention that some "pollutants", like the bacteria or some chemical compounds, have a "life" of their own, in the sense that their presence can change in accordance with specific biological or chemical processes occurring and enhanced by the contact with water, giving rise to other substances through *reactions* or *mutations*. These will not be considered in

M. Benedini and G. Tsakiris, *Water Quality Modelling for Rivers and Streams*,
Water Science and Technology Library 70, DOI 10.1007/978-94-007-5509-3_3,
© Springer Science+Business Media Dordrecht 2013

Fig. 3.1 Interpretation of the
presence of pollutant particles
(*black dots*) and water
particles (*grey dots*) in the
elementary volume of the
water body

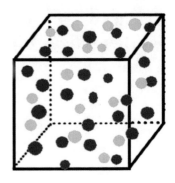

relation with their intrinsic nature but as their ultimate effect of changing the
measurable quantity of the pollutant.

The pollutants of this kind are *nonconservative* (or *nonpersistent*). On the
contrary, the pollutants that do not undergo similar alterations are *conservative*
(or *persistent*).

The fundamental aspects of pollution transport in water are described in the
following paragraphs. For the sake of clarity, it is better to consider first the case of
conservative pollutants.

As mentioned, a pollutant in water is quantified by its *concentration*, expressed
in terms of mass per unit volume, and its dimensions being $[ML^{-3}]$.

In case the pollutant presence is in form of chemical solution, the concentration
can be expressed in terms of *mole* of pollutant per unit water volume, where
1 mole = 6.02×10^{23} molecules of the chemical compound.

If Φ is an elementary water volume and M is the mass of pollutant inside the
same volume at the time t_0, the concentration can be expressed as

$$C = \frac{M}{\Phi} \tag{3.1}$$

In a conceptual scheme (Fig. 3.1), the presence of a polluting substance in water
can be thought as a mixture of pollutant and water microscopic particles, each one is
in a different number, according to the concentration, which is determined by the
number of pollutant particles contained in the elementary volume.

As already mentioned, due to various phenomena connected with the intrinsic
behaviour of the water body, and to the human intervention as well, the concentration varies from point to point and from one instant to another. In other words, it is a
function of both time and space:

$$c = c(x, y, z; t)$$

Considering two contiguous points P_1 and P_2 in the water volume, it is possible
that, during a certain time interval, the pollutant is *transported* from the former to
the latter, and therefore, the concentration is decreased in P_1 and increases in P_2.

Fig. 3.2 The elementary
volume around the generic
point P

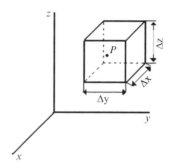

As illustrated in Fig. 3.2, considering now an elementary volume of water
$\Delta x \cdot \Delta y \cdot \Delta z$ around the point P, in which C_t is the average pollutant concentration
at the time t, the pollutant mass in the volume can be expressed as

$$C_t \Delta x \Delta y \Delta z \tag{3.2}$$

At the time $t + \Delta t$, the average concentration is $C_{t+\Delta t}$ and the corresponding
mass is $C_{t+\Delta t} \Delta x \Delta y \Delta z$. Therefore, during the time interval Δt, the mass has
changed of

$$(C_{t+\Delta t} - C_t) \Delta x \Delta y \Delta z \tag{3.3}$$

There are various forms and mechanisms that govern the pollution transport in
water.

3.2 The Advection Transport

The first and the most important mechanism of pollutant transport from one place to
another by fluid flow is the *advection* (or *convection*) (Gulliver 2007). The conceptual aspect of such transport can be illustrated in Fig. 3.3, in which the water
particles are described as elementary "trucks" conveying the pollutant particles
along the relevant streamlines.

The advection transport depends on *water velocity* at the point $P(x, y, z)$
considered, namely,

$$v = v(x, y, z; t) \tag{3.4}$$

which, in turn, depends on time t and has components v_x, v_y and v_z in the orthogonal
system of coordinates in the directions x, y and z, respectively. Therefore, the water
velocity is a vector.

Fig. 3.3 Conceptualisation
of the advection transport: the
water particles (the "trucks"),
moving along the streamlines
(*solid lines*), convey the
pollutant particles (the
"balls")

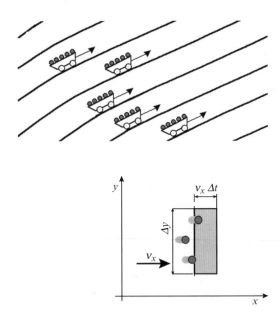

Fig. 3.4 Interpretation of the
advection transport in a two-
dimensional case ($\Delta z = 1$)

To formally interpret the advection transport, according to Fig. 3.4 (which refers, for simplicity, to a two-dimensional case with $\Delta z = 1$), the pollutant particles are transported by water moving with velocity v_x through the elementary area $\Delta y \, \Delta z$. If C is the pollutant concentration, the pollutant mass crossing this area during the time interval Δt can be quantified as

$$C\Delta y \Delta z \, v_x \, \Delta t \tag{3.5}$$

3.3   The Dispersion Transport: Fick Law

The pollutant concentration in the water volume can vary even in the absence of motion, because other facts can occur in a way not necessarily dependent on the migration of water particles. A physically meaningful mathematical description of diffusion that is based on the analogy to heat conduction is the first Fick law established by Fick (1855; Gulliver 2007).

This process is called *dispersion* (sometimes also *diffusion*) and is due to the variability of concentration inside the water. In order to formally express this process, two new fundamental terms are introduced, namely:

- The *pollutant flux*, that is, the mass crossing the unit area in unit time

$$J = J \, (x, y, z; t) \tag{3.6}$$

with dimensions $[\mathrm{ML}^{-2} \, \mathrm{T}^{-1}]$

Fig. 3.5 Conceptual aspect
of the dispersion pollutant
transport

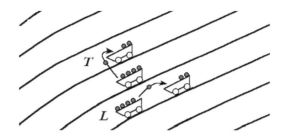

- The *concentration gradient*, that is, the difference of concentration in a given direction between two points located at an infinitesimal distance apart

$$\frac{dC}{dx} = \mathrm{grad}C \tag{3.7}$$

with dimensions $[\mathrm{ML}^{-4}]$

Both the above terms depend on the space coordinates and are characterised by the geometrical direction considered. Therefore, they can be expressed as vectors.

The dispersion process is explained considering that a difference of concentration between two points at an infinitesimal distance from each other will cause the migration of pollutant particles from the point of higher to that of lower concentration, in order to reach a general equilibrium, in accordance with an overall principle of nature. This can be explained through the elementary interpretation illustrated in Fig. 3.5, where the "trucks" are supposed to carry a different amount of pollutant particles. The need to reach the general equilibrium induces the particles to migrate to the nearest "truck", both following the streamline (L = *longitudinal dispersion*) or across (T = *transversal dispersion*).

The mechanism is currently interpreted by means of *first Fick law*, which states that

$$\vec{J} = -E \; \overrightarrow{\mathrm{grad}} \; C \tag{3.8}$$

where the factor E is the *dispersion coefficient* which is characteristic of the liquid field, with dimension $[\mathrm{L}^2\,\mathrm{T}^{-1}]$. The negative sign means that, as in Fig. 3.6, the flux (which is in accordance with the positive x-direction) is from the side of higher to that of lesser concentration.

In the preceding statements, E is assumed to vary from point to point:

$$E = E(x, y, z) \tag{3.9}$$

but in many cases, it may be considered constant for all the liquid volume (*homogeneous fluid*). Moreover, it is generally assumed to vary according to the direction considered in the fluid volume (in such a case, it is mathematically defined as a "second-order tensor") and the fluid is *anisotropic*; if, conversely, it can be assumed

Fig. 3.6 Interpretation of the dispersion transport in a two-dimensional ($\Delta z = 1$) stream

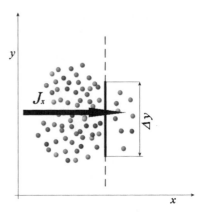

independent of any direction, the fluid is *isotropic*. Water in the surface water bodies is considered homogeneous and isotropic.

As in Fig. 3.6, following the application of Fick law to an elementary area $\Delta y\,\Delta z$ in the water volume, the mass crossing the area during the time interval Δt is

$$J_x\Delta y\Delta z\Delta t = -\left(E\frac{\partial C}{\partial x}\right)_x \Delta y\Delta z\Delta t \tag{3.10}$$

3.4   External Contributions

Another way of changing pollutant concentrations in water is due to *local injection* (*source*) from outside the water volume, or *local subtraction* (*sink*), or else due to any other cause that may occur in leaching or absorbing pollutants at the bottom, the wall or at the free surface of the water body taken into consideration. This can be expressed in a general form by means of a term $\pm S(x, y, z; t)$ having dimension $[ML^{-3}\,T^{-1}]$. The sign (+) means injection and the (−) subtraction.

The way sinks and sources act will be better explained in Chap. 7.

3.5   In-Water Transformations

The pollutants already defined nonconservative have the peculiarity that, once in contact with the water of the water body where they are immersed and in presence of other active substances, they enter chemical reactions able to transform their original molecular composition, giving rise to different compounds. Transformations can occur also for bacteria and other living organisms, which are characterised by a form of "life". The contact with water and some chemicals

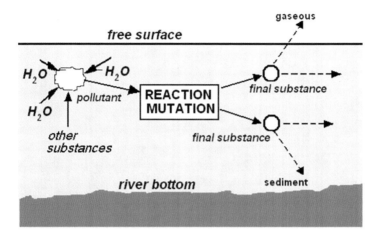

Fig. 3.7 Behaviour of nonconservative pollutants in contact with water

present in the water volume can foster their growth and eventually their death, with the final appearance of different substances of vital or mineral nature. Moreover, the new substances and compounds can behave in a way that is different from that of the original pollutant and remain in the water volume or settle at the bottom in form of sediments or volatilise through the free surface in form of gas, as illustrated in Fig. 3.7.

The ultimate effect of such transformations is the variation of the original pollutant concentration, following a mechanism that is assumed to be a function of time and be proportional to the initial concentration ("the higher is the original concentration, the greater will be the amount transformed"). This assumption facilitates the mathematical interpretation of these processes.

These complex phenomena will be better examined in the following chapters.

References

Fick AE (1855) Poggendorff's Annelen der Physic 94:59

Foster I, Gurnell A, Webb B (eds) (1995) Sediment and water quality in river sediments. Wiley, Chichester

Gray N (2008) Drinking water quality. Problems and solutions, 2nd edn. Cambridge University Press, Cambridge

Gulliver J (2007) Introduction to chemical transport in the environment. Cambridge University Press, New York

Trudgill S, Walling D, Webb B (1999) Water quality. Processes and policy. Wiley, Chichester

Quevauviller P (2002) Quality assurance for water analysis, Water quality measurement series. Wiley, Chichester

Chapter 4
Fundamental Expressions

Abstract Considering a water elementary volume, the forms and mechanisms of pollutant transport allow for the formulation of an overall balance of the pollutant mass. In an infinitesimal volume (control volume), the fundamental differential equation of pollutant transport can be written, the integration of which can give at any instant the concentration of the pollutant at any point in the water body. For the application of the fundamental differential equation, the velocity field should be determined. For this purpose, river hydraulics provides answers produced through complex calculations. Simplified procedures are also proposed, able to speed up the calculation reaching sufficiently accurate results.

4.1 The General Equation of Pollutant Transport in Water

It is now possible to formally describe mathematically how the pollutant mass varies at a certain point of the water volume, due to the various processes previously described.

For the sake of simplicity, at an instant t, instead of the elementary control volume of water $\Delta x\,\Delta y\,\Delta z$, a two-dimensional volume is assumed in the x-y plain with a motion parallel to the x-direction having velocity v_x.

Due to advection, in the time interval Δt, the mass of the pollutant entering the volume through the face $\Delta y\,\Delta z$ at location x is

$$\Delta y\Delta z\, v_x C_x \Delta t \tag{4.1a}$$

and that going out through the face $\Delta y\,\Delta z$ at location $x + \Delta x$ is

$$\Delta y\Delta z\, v_{x+\Delta x}C_{x+\Delta x}\Delta t \tag{4.1b}$$

M. Benedini and G. Tsakiris, *Water Quality Modelling for Rivers and Streams*,
Water Science and Technology Library 70, DOI 10.1007/978-94-007-5509-3_4,
© Springer Science+Business Media Dordrecht 2013

The difference between the quantities (4.1b) and (4.1a) is the mass variation in the Δt interval:

$$\Delta y \Delta z \left(v_{x+\Delta x} C_{x+\Delta x} - v_x C_x \right) \Delta t \tag{4.2}$$

If this term is positive, the outflow of pollutant mass is greater than the inflow. Therefore, during the interval Δt, the pollutant mass in the elementary volume decreases, and its balance is

$$- \Delta x \Delta y \Delta z (C_{t+\Delta t} - C_t) = \Delta y \Delta z (v_{x+\Delta x} C_{x+\Delta x} - v_x C_x) \Delta t \tag{4.3a}$$

or, after dividing by $\Delta x \, \Delta y \, \Delta z \, \Delta t$,

$$- \frac{(C_{t+\Delta t} - C_t)}{\Delta t} = \frac{(v_{x+\Delta x} C_{x+\Delta x} - v_x C_x)}{\Delta x} \tag{4.3b}$$

which, if the size of the elementary volume is infinitesimal (i.e. $\Delta x \to 0$, $\Delta y \to 0$ and $\Delta z \to 0$), in an infinitesimal time interval ($\Delta t \to 0$), becomes

$$- \left(\frac{\partial C}{\partial t} \right)_{adv} = \frac{\partial (v_x C)}{\partial x} \tag{4.4}$$

with the minus sign $(-)$ in the left-hand side.

In the same Δt interval, the elementary volume undergoes the effect of dispersion. The mass inflow through the face $\Delta y \, \Delta z$ at location x is

$$- \left(E \frac{\partial C}{\partial x} \right)_x \Delta y \Delta z \Delta t \tag{4.5a}$$

and that going out through the face $\Delta y \, \Delta z$ at $x + \Delta x$

$$- \left(E \frac{\partial C}{\partial x} \right)_{x+\Delta x} \Delta y \Delta z \Delta t \tag{4.5b}$$

The difference between (4.5b) and (4.5a) is the balance of pollutant mass in the elementary volume

$$- E \left[\left(\frac{\partial C}{\partial x} \right)_{x+\Delta x} - \left(\frac{\partial C}{\partial x} \right)_x \right] \Delta y \Delta z \Delta t \tag{4.6}$$

Also, in this transformation, if the term between the square brackets is positive, the mass going out is greater than that entering and the total mass in the elementary volume decreases. With the same considerations performed above for the advection, the pollutant mass balance is

$$- \Delta x \Delta y \Delta z (C_{t+\Delta t} - C_t) = -E \left[\left(\frac{\partial C}{\partial x} \right)_{x+\Delta x} - \left(\frac{\partial C}{\partial x} \right)_x \right] \Delta y \Delta z \Delta t$$

leading to

$$\left(\frac{\partial C}{\partial t} \right)_{disp} = E \frac{\partial^2 C}{\partial x^2} \qquad (4.7)$$

This expression is known as the *second Fick Law*.

To complete this picture, the local injection or subtraction, in terms of the pollutant mass, can be considered as

$$S \ \Delta x \Delta y \Delta z \Delta t$$

in which S is the concentration of the pollutant of injection or subtraction. The balance of pollutant mass becomes

$$\Delta x \Delta y \Delta z \ (C_{t+\Delta t} - C_t) = S \ \Delta x \Delta y \Delta z \Delta t \qquad (4.8)$$

If $C_{t+\Delta t} > C_t$, the mass in the elementary volume increases during the Δt interval and the left-hand side of (4.8) is positive: S is also positive and acts as a pollution source. Therefore, for the infinitesimal control volume,

$$\left(\frac{\partial C}{\partial t} \right)_{conc} = S \qquad (4.9)$$

The combined effect of advection, dispersion and local injection or subtraction is the sum of the expressions (4.4), (4.7) and (4.9) according to the *principle of superposition*:

$$\frac{\partial C}{\partial t} = \left(\frac{\partial C}{\partial t} \right)_{adv} + \left(\frac{\partial C}{\partial t} \right)_{disp} + \left(\frac{\partial C}{\partial t} \right)_{conc} \qquad (4.10)$$

Taking into consideration the relevant signs in the right-hand side, the mass balance in the infinitesimal elementary volume along the x-direction is

$$\frac{\partial C}{\partial t} = -\frac{\partial v_x C}{\partial x} + E \frac{\partial^2 C}{\partial x^2} + S \qquad (4.11)$$

Extending the above considerations to the three dimensions, Eq. (4.11) becomes

$$\frac{\partial C}{\partial t} + \left(\frac{\partial v_x C}{\partial x} + \frac{\partial v_y C}{\partial y} + \frac{\partial v_z C}{\partial z} \right) = E \left(\frac{\partial^2 C}{\partial x^2} + \frac{\partial^2 C}{\partial y^2} + \frac{\partial^2 C}{\partial z^2} \right) + S \qquad (4.12)$$

where S is now considered at the point P around which the elementary volume is assumed. A positive value of S means that there is an increase of the pollutant mass in the elementary volume.

Equation (4.12) is the fundamental differential equation of the transport of a conservative pollutant. After its integration, within appropriate initial and boundary conditions, Eq. (4.12) gives the pollutant concentration at any point in the water volume at any instant.

4.2   Nonconservative Pollutants

As mentioned in Chap. 3, the variation of pollutant mass in contact with water can be the result of chemical reactions or biological evolutions, by means of which the original pollutant particle is modified and other substances appear. These reactions are conditioned by the water temperature and the other chemicals present in the water which often contribute to build up the final substance.

To evaluate the water quality, in the context adopted so far, there is no need of interpreting the inner mechanisms of these chemical or biological transformations, which are indeed very complex and difficult to be understood. It is conversely essential to consider the change of concentration, which is the ultimate effect of the mechanisms acting in the water volume.

It is customarily accepted that the mass variation of such pollutants in the elementary volume of water and in the unit time depends on the initial pollutant concentration. The greater the amount of pollutant present in the water, the greater is the transformed quantity.

This assumption is generally described by the expression:

$$-\frac{\partial C}{\partial t} = k(C - C_0)^n \qquad (4.13)$$

in which k and n are appropriate terms, while C_0 is a reference concentration. All these terms are characteristic of the pollutant considered. The negative sign in (4.13) means that if the right-hand side is positive, the pollutant mass in the infinitesimal elementary volume decreases: the effect is, therefore, a pollutant abatement known as *decay*.

Expression (4.13) is known in chemistry as "*n-order kinetic*". Very frequent is the case of $n = 1$ that gives the "*linear*" or "*first-order kinetic*". Frequent is also the case in which $C_0 = 0$. The dimension of k becomes, therefore, $[T^{-1}]$; it can be called simply *reaction coefficient*, but other definitions are common, as it is explained in the following chapters.

Generally speaking, the terms k and n can vary with respect to time, but due to their complexity and the difficulty for their measurement, they are often considered constant.

In a water body, these reactions or transformations occur in conjunction with all the other mechanisms responsible for changing the pollutant concentration, already described in the preceding paragraphs. In the assumption that the cumulative effect is the sum of the single ones, the behaviour of a nonconservative pollutant in the elementary volume of water can be described by adding expression (4.13) to expression (4.12).

Therefore, the general differential equation of a nonconservative pollutant transport may be written as

$$\frac{\partial C}{\partial t} + \left(\frac{\partial v_x C}{\partial x} + \frac{\partial v_y C}{\partial y} + \frac{\partial v_z C}{\partial z} \right) = E \left(\frac{\partial^2 C}{\partial x^2} + \frac{\partial^2 C}{\partial y^2} + \frac{\partial^2 C}{\partial z^2} \right) + S - kC \qquad (4.14)$$

The integration of this equation provides the capability for the detection of the behaviour of the nonconservative pollutant in a water body.

4.3   Combined Processes

A very common process is the alteration of the organic matter present in the water in form of solution and suspension, due to the activity of multispecies microbial populations. The bacteria contribute to the transformation of the biodegradable substance into more stable compounds using the oxygen present in water. The process is normally referred to in terms of concentration of *biochemical oxygen demand* (BOD), which is the amount of oxygen necessary to fulfil all the various stages of the transformation. This matter will be treated with more details in Chap. 7.

The BOD is a reliable indicator of the presence of both bacteria and transformable organic load discharged from a polluting source. It is, therefore, very useful for the assessment of the pollution level reached by the water body. Control of BOD is essential in water quality monitoring.

As the oxygen to fulfil these processes is provided by that present in the water, the concentration of *dissolved oxygen* (DO) is also a useful indicator of the water quality; DO takes into account the fact that a certain amount of oxygen is also necessary for other chemical processes.

As a matter of fact, the higher the BOD concentration is in a water body, the lower is the concentration of DO. Moreover, as a consequence of this process, as the amount of bacteria increases and the organic matter decreases, the greater is the quantity of oxygen unused and left in water in form of DO. The oxidation process causes the *decay* of the BOD initially present in the water body.

This mutual behaviour of BOD and DO can be considered in a global form, without entering into the inner chemical and biological processes but referring only to their presence in the water body.

As it will be explained in following chapters, the DO concentration can be expressed as *deficit from saturation*. If σ is the DO concentration and c is that of BOD, this concept can be expressed by

$$\frac{\partial \sigma}{\partial t} = -K_a \sigma + Kc \tag{4.15}$$

in which K_a is normally defined *reaeration coefficient* and K *deoxygenation coefficient*.

In the water body, the described process must be combined with those of advection, dispersion and local injection (or subtraction). Adopting, as usual, a linear combination of the various effects, the mutual behaviour can be examined by means of the following two equations:

$$\frac{\partial c}{\partial t} + \left(\frac{\partial v_x c}{\partial x} + \frac{\partial v_y c}{\partial y} + \frac{\partial v_z c}{\partial z} \right) = E \left(\frac{\partial^2 c}{\partial x^2} + \frac{\partial^2 c}{\partial y^2} + \frac{\partial^2 c}{\partial z^2} \right) + S_{BOD} - Kc \tag{4.16a}$$

and

$$\frac{\partial \sigma}{\partial t} + \left(\frac{\partial v_x \sigma}{\partial x} + \frac{\partial v_y \sigma}{\partial y} + \frac{\partial v_z \sigma}{\partial z} \right) = E \left(\frac{\partial^2 \sigma}{\partial x^2} + \frac{\partial^2 \sigma}{\partial y^2} + \frac{\partial^2 \sigma}{\partial z^2} \right) + S_{DO} - K_a \sigma + Kc \tag{4.16b}$$

This system of differential equations describes a two-component process. In a similar way, a multicomponent process, involving several different substances interacting with each other, can be described. The process is mathematically interpreted by means of coefficients of mutual interaction, in the equations relevant to each pollutant.

If C_p is the concentration of pollutant p and C_q that of pollutant q, the problem is interpreted by means of a system of differential equations of the type (in the x-dimension):

$$\frac{\partial C_p}{\partial t} + \frac{\partial v_x C}{\partial x} = E \frac{\partial^2 C}{\partial x^2} \pm S_p \mp k_p C_p \mp k_q C_q \tag{4.17}$$

in which the term $k_p C_p$ (with negative sign) accounts for the decay of p-th pollutant, while $k_q C_q$ accounts for the increment of q-th pollutant due to the transformation of p-th pollutant. Very common is the case of nitrogen compounds, which are produced from the discharge of pollutants which are modified in contact with water.

All these aspects will be examined with more details in Chap. 7.

4.4 Hydrodynamic Aspects

As known, fluid flow can be described by Navier-Stokes equations, and hence, the velocities can be determined in the entire computational field. These equations are in three-dimensional (3D) form and are based on mass conservation and the second Newton law for momentum conservation in a control volume. Mathematically the mass and momentum equations can be written as follows (Chaudhry 2008):

Mass conservation

$$\frac{\partial u}{\partial x} + \frac{\partial v}{\partial y} + \frac{\partial w}{\partial z} = 0 \tag{4.18}$$

Momentum conservation

$$\frac{\partial u}{\partial t} + v_x \frac{\partial u}{\partial x} + v_y \frac{\partial u}{\partial y} + v_z \frac{\partial u}{\partial z} = g_x - \frac{1}{\rho} \frac{\partial p}{\partial x} + \frac{\mu}{\rho} \nabla^2 u$$

$$\frac{\partial v}{\partial t} + v_x \frac{\partial v}{\partial x} + v_y \frac{\partial v}{\partial y} + v_z \frac{\partial v}{\partial z} = g_y - \frac{1}{\rho} \frac{\partial p}{\partial y} + \frac{\mu}{\rho} \nabla^2 v$$

$$\frac{\partial w}{\partial t} + v_x \frac{\partial w}{\partial x} + v_y \frac{\partial w}{\partial y} + v_z \frac{\partial w}{\partial z} = g_z - \frac{1}{\rho} \frac{\partial p}{\partial z} + \frac{\mu}{\rho} \nabla^2 w \tag{4.19}$$

where

$$\nabla^2 = \frac{\partial^2}{\partial x^2} + \frac{\partial^2}{\partial y^2} + \frac{\partial^2}{\partial z^2}$$

and

u, v, w = velocity components at x-, y-, z-direction $[LT^{-1}]$
g = acceleration of gravity $[LT^{-2}]$
μ = dynamic viscosity $[ML^{-1} T^{-1}]$
p = pressure $[ML^{-2} T^{-2}]$
ρ = fluid density $[ML^{-3}]$

In case the vertical component of the flow velocity is negligible in comparison with the retrospective transverse or longitudinal components (river flow, coastal areas), the Navier-Stokes equations can be inferred integrating on one or two dimensions (1D or 2D). The most known integration are the *Shallow Water Equations* (SWE) in one or two dimensions (1D-SWE or 2D-SWE), known also as Saint-Venant equations (Barré de Saint-Venant 1871), which are valid under the following conditions (Abbott 1979):

- Water is incompressible and homogeneous.
- Velocity components in the vertical direction are negligible.
- Pressure distribution is hydrostatic in the vertical direction.

- Bottom slope is small (water depth is the same if it is measured normally or vertically).
- There are no discontinuities in the flow field.
- There is no source term such as rainfall or sink term such as evaporation.
- Friction terms (viscosity, bottom and free surface friction) can be simulated by semiempirical expressions (Manning, Chèzy, etc.) of steady flow.

In 2D-SWE form, the longitudinal velocity is of the same order as the retrospective transverse component. Equations 2D-SWE are suitable for the simulation of phenomena such as flood propagation in mild terrain and water flow in coastal areas. With suitable modifications in Navier-Stokes equations, the 2D-SWE are shown below in law conservation form (suitable for numerical methods):

$$\frac{\partial W}{\partial t} + \frac{\partial F}{\partial x} + \frac{\partial G}{\partial y} = D \tag{4.20}$$

where

$$W = \begin{vmatrix} h \\ uh \\ vh \end{vmatrix} \quad F = \begin{vmatrix} uh \\ u^2h + \frac{gh^2}{2} \\ uvh \end{vmatrix} \quad G = \begin{vmatrix} vh \\ uvh \\ v^2h + \frac{gh^2}{2} \end{vmatrix} \quad D = \begin{vmatrix} 0 \\ gh(S_0^x - S_f^x) \\ gh(S_0^y - S_f^y) \end{vmatrix}$$

and

h = water depth [L]
u = flow velocity at horizontal direction x [LT^{-1}]
v = flow velocity at horizontal direction y [LT^{-1}]
g = acceleration of gravity [LT^{-2}]
S_0^x = bottom slope in horizontal direction x [−]
S_0^y = bottom slope in horizontal direction y [−]
S_f^x = slope of energy line in horizontal direction x [−]
S_f^y = slope of energy line in horizontal direction y [−]

The above terms of energy slope can be determined empirically:

$$S_f^x = \frac{n^2 u \sqrt{u^2 + v^2}}{h^{\frac{4}{3}}} \tag{4.21a}$$

$$S_f^y = \frac{n^2 v \sqrt{u^2 + v^2}}{h^{\frac{4}{3}}} \tag{4.21b}$$

where n is the *Manning roughness coefficient* [L$^{-1/3}$ T].

   In cases when the transverse velocity components of fluid flow are much smaller than the longitudinal ones (stream flow), they can be considered negligible. Due to this simplification, it is possible to write the 1D-SWE based on mass and momentum conservation in a form that is suitable for the numerical methods:

$$\frac{\partial W}{\partial t} + \frac{\partial F}{\partial x} = D \tag{4.22}$$

where

$$W = \begin{vmatrix} A \\ UA \end{vmatrix} \quad F = \begin{vmatrix} UA \\ U^2A + gA\bar{y} \end{vmatrix} \quad D = \begin{vmatrix} q \\ gA(S_0 - S_f) \end{vmatrix}$$

and

A = area of wetted cross section [L^2]
U = water velocity [LT^{-1}]
q = lateral inflow per unit length of the river [M^2T^{-1}]
g = acceleration due to gravity [LT^{-2}]
$\bar{y}$ = distance between the centroid of the cross section and the water surface [L]
S_0 = bottom slope [$-$]
S_f = slope of energy line [$-$]

   The component S_f can be calculated as

$$S_f = \frac{n^2 U^2}{R^{\frac{4}{3}}} \tag{4.23}$$

where n is the Manning roughness coefficient and R the hydraulic radius [L].

4.5   A Simplified Interpretation

In a river course where the transverse velocity components play a secondary role, the 1D-SWE can describe the phenomena satisfactorily and can be written in the form

$$\frac{\partial h}{\partial t} + \frac{\partial (Uh)}{\partial x} = q \tag{4.24a}$$

$$\left(\frac{\partial U}{\partial t} + U \frac{\partial U}{\partial x} \right) + g \frac{\partial h}{\partial x} = g(S_0 - S_f) \tag{4.24b}$$

where, at the x longitudinal location, h is the water depth [L].

Equation (4.24a) expresses the continuity of the volume enclosed in an elementary length extending all over the cross section. Equation (4.24b) gives the momentum balance; its first term is the *local acceleration,* and the second term is the *convective acceleration.* The slope of energy line takes into consideration the resistance encountered in the water flow along the bottom and banks of the river.

According to this approach, the dynamic characteristics of the river (water velocity and water depth) are dependent on both longitudinal (x) and time (t) variables. In this way, the problem can consider also the nonstationary phenomena, like the propagation of waves, very important for the pollution transport.

There is no analytical solution of SWE, and hence, the solution can be achieved by adopting appropriate mathematical procedures in numerical methods, now facilitated by the availability of suitable software packages for computer applications. The numerical methods can be categorised in *finite difference method* (FDM), *finite element method* (FEM) and *finite volume method* (FVM). Several attempts have been made to handle SWE jointly with those of pollution transport, even though by means of separate procedures (Schaffranek 1998), in order to have a better view of the entire phenomenon. Ad hoc software has been also proposed to speed up the calculations (Modenesi et al. 2004). It should be noticed that good results can be achieved in spite of simplifying approximations of the channel geometry and behaviour.

In the majority of practical cases, the problems of water quality in a river are satisfactorily approached in *steady conditions*, not dependent on time. For any stretch, the well-known equations of uniform flow can be valid. Among these relations, the *Manning formula* is the most frequently used, according to which the mean water velocity U is calculated as follows:

$$U = \frac{1}{n} R^{\frac{2}{3}} S_0^{\frac{1}{2}} \qquad (4.25)$$

or for streams with very wide cross section (in which $R \approx h$)

$$U = \frac{1}{n} h^{\frac{2}{3}} S_0^{\frac{1}{2}} \qquad (4.25a)$$

The most common values of Manning roughness coefficient, n, for rivers are presented in Table 4.1.

Dealing with the water quality problems in a river, a thorough application of the procedures previously described entails normally wearisome steps and requests data not always available. Moreover, it concerns an order of approximation that cannot be maintained with the real appreciation of the problem. In order to simplify the approach, once the average velocity is obtained, some authors propose to infer the fundamental terms of the river by means of empirical relations, able to maintain the results of the calculation within acceptable limits of approximation.

Table 4.1 Manning roughness coefficient (m$^{-1/3}$ s) for rivers

Type of surface	State of the channel			
	Very good	Good	Bad	Very bad
Earth, straight and clean	0.017	0.020	0.022	0.025
Earth, dragged	0.025	0.027	0.030	0.033
Rock, smooth	0.025	0.030	0.033	0.035
Rock, irregular	0.035	0.040	0.045	0.048
Weeds, pebbles, winding	0.025	0.030	0.035	0.045
Earth, with stones	0.028	0.030	0.040	0.050

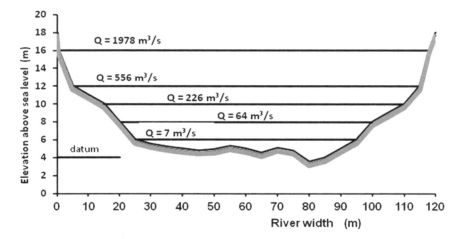

Fig. 4.1 Free surface elevation for the corresponding flow values in a river cross section

For a well-defined river cross section, the flow Q, the average velocity U, the river depth h and its width B can be connected by very simple relations, as explained in the following example:

In all the rivers, there are reaches that are sufficiently straight and uniform, where the flow is related to the water level in a unique way, as illustrated in Fig. 4.1. In this figure, the water levels are shown in relation to the flow in a cross section of an urban reach of Tiber River in Italy.

A relationship between flow and water depth, known as *rating curve*, can be often described by a power function such as

$$h = aQ^{\beta} \tag{4.26}$$

where h is in metres (m) and Q in m^3/s.

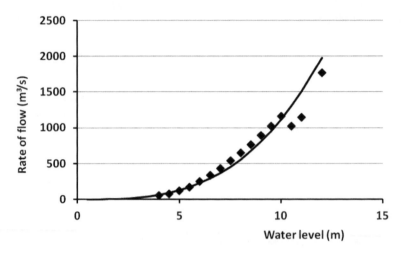

Fig. 4.2 Rating curve of the river cross section shown in Fig. 4.1

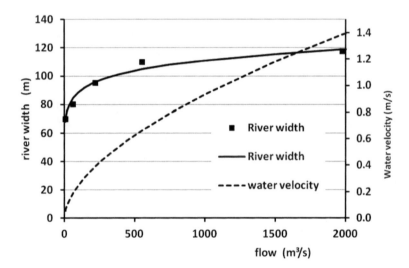

Fig. 4.3 River width vs. flow (*solid line*) and velocity flow (*dotted line*) in the cross section of Fig. 4.3

The values of the parameters α and β can be determined by regression using the pairs (flow, water depth) after a logarithmic transformation. For the particular case examined (Fig. 4.1), $\alpha = 1.061$ and $\beta = 0.3195$.

In the cross section, at the water level determined by a selected flow, there is a corresponding unique value of the river width B, as indicated in Fig. 4.3. It is now possible to correlate the width to the flow, obtaining also a simple expression:

$$B = \chi Q^f \tag{4.27}$$

In the case examined, $\chi = 56.03$ and $f = 0.099$, with the same units of B and Q as above.

Considering that $Q = BhU$, after combining (4.26) with (4.27), replacing in the exponent the quantity $1-f-\beta$ with **b** and putting $(\chi\alpha)^{-1} = \mathbf{a}$, the velocity can be written as a function of flow by means of the following expression:

$$U = \mathbf{a}Q^{\mathbf{b}} \tag{4.28}$$

For the case presented above, $\mathbf{a} = 0.0168$ and $\mathbf{b} = 0.5815$.

The example just described was originally proposed by Leopold and Maddock (1953). As noticed, this simplified approach is very useful and can simplify the determination of the hydraulic field in a river. Care should be taken when dealing with very low or very high flows, because expressions (4.26), (4.27) and (4.28) could be less reliable. In fact, normally, the extreme limbs of the rating curve are obtained after an extrapolation of values of the middle range, which are more frequent and can be experimentally confirmed. On the opposite, it is not always possible to obtain significant measurements in a river during the high flow event or in drought period.

It is now worthy to notice that the procedure of the preceding paragraphs is especially developed for simplifying the application of the water quality models in which the hydraulic and hydrological terms are "exogenous", and are used in the algorithm as parameters or constant values. Several attempts have been made for the development of combined models able to treat both the hydraulic and water quality aspects (Koussis 1983; Koussis and Saenz 1983) in a way that the river characteristics can be evaluated with much precision and the pollution transport is simulated in a more realistic environment.

References

Abbott MB (1979) Computational hydraulics. Pitman Publishing Ltd., London, p 324

Barré de Saint-Venant AJC (1871) Théorie du mouvement non-permanent des eaux avec application aux crues des rivières et à l' introduction des marées dans leur lit. C R Acad Sc Paris 73:147–154

Chaudhry MH (2008) Open channel flow, 2nd edn. Springer, New York, p 523

Koussis AD (1983) Unified theory for flood and pollution routing. J Hydraul Eng ASCE 109 (12):1652–1664

Koussis AD, Saenz MA (1983) Pollution routing in streams. J Hydraul Eng ASCE 109 (12):1636–1651

Leopold LB, Maddock T (1953) The hydraulic geometry channels and some physiographic implications, Geological Survey professional paper 252. U.S. Government Printing Office, Washington DC, USA

Modenesi K, Furlan LT, Guirardello R, Nunez JR (2004) A CFD model for pollutant dispersion in rivers. Braz J Chem Eng 21(4); Sao Paulo, Brazil

Schaffranek RW (1998) Simulation model for open-channel flow and transport. In: Proceedings of the first federal interagency hydrologic modeling conference, Las Vegas, NV (1), pp 103–110

Chapter 5
Dispersion in Rivers and Streams

Abstract The water quality problems of rivers and streams are controlled by the natural behaviour of the water body, which is interpreted by means of proper terms and expressions of free surface hydraulics. Pollution transport is due primarily to advection, but there are many situations in which dispersion plays an important role and cannot be neglected. In the mathematical models, the effect of dispersion is accounted by means of the dispersion coefficient, for the evaluation of which several procedures are proposed, supported by experimental studies.

5.1 The Importance of Dispersion

Dispersion refers to the spreading and mixing of pollutant mass and includes the combined effects of turbulent mixing, molecular diffusion and mixing due to vertical and transverse shear (Singh and Beck 2003). This process changes the concentration of any substance which is present in the water body. Moreover, it is one of the most important factors for evaluating the pollutant behaviour in a river or stream (Kim et al. 2007).

The dispersion process in rivers is often less important in the transport of pollutants, when compared to the advection, which predominates due to the relatively high water velocity. According to Zhen-Gang (2008), the effect of dispersion may be ignored in analysing a continuous pollutant load in a river. Nevertheless, in some large rivers, there are local conditions or occasions in which the velocity becomes very low, and the pollution transport is performed to a great extent by dispersion. This occurs, for instance, in the lowest downstream stretches of a river, where it approaches the estuary and the sea, and the water velocity values sometimes are close to zero. Consequently, considering the transport as due to advection only is not correct (Chaiwiwatworakul et al. 2005; Antonopoulos and Papamichail 1991; Riahi-Madvar and Ayyoubzadeh 2010; Deng 2002; Deng and Jung 2007; Ki-Chul et al. 2011).

The evaluation of the dispersion coefficient E is still controversial, because it cannot be measured directly but can be appreciated only after the application of

M. Benedini and G. Tsakiris, *Water Quality Modelling for Rivers and Streams*,
Water Science and Technology Library 70, DOI 10.1007/978-94-007-5509-3_5,
© Springer Science+Business Media Dordrecht 2013

some expressions in which it is related to known terms that can be measured. Theoretically, the correct way of its evaluation should be the application of the equations of pollution transport to a case in which all the involved terms are known, with the exception of E, which becomes the unique unknown of the problem.

In practice, there are several reasons of incertitude, due to the real conditions of the river, and consequently, there is always the need of adopting some empirical adjustments. Another difficult task is to make comparable the values obtained in different cases and using different tools.

The inner nature of the dispersion phenomena requires, first of all, to split the dispersion coefficient in two components, namely, the *coefficient of longitudinal dispersion*, E_L, and the *coefficient of transverse dispersion* E_{tr} (Baek and Seo 2005; Sooky 1969; Fukuoka and Sayre 1973). These considerations are in line with the general theory already explained in Chap. 4, but should be adapted to the specific case of a river (Rutherford 1994).

Numerous research efforts have pointed out the dependence of the coefficient on the real characteristics of the flowing water, recalling the main principles of fluid dynamics (Liu 1977; Nordin and Troutman 1980; Ranga-Raju 1987; Guymer and West 1992; Seo and Baek 2004; Czernuszenko 1987; Shen 1980; Givehchi et al. 2009). Other accurate investigations (Tayfur and Singh 2005; Iwasa and Aya 1991; Bogle 1997; Baek and Seo 2005) have applied the most up-to-date resorts of the advanced scientific output enhanced by the technological progress.

Longitudinal dispersion, in the direction of water flow, acts along the river stretches, while the transverse dispersion occurs in the cross section, both horizontally and vertically. In a river, the longitudinal dispersion is predominant, and the transverse dispersion is significant only in the case of a stretch having very low velocity (Carr and Rehmann 2005; Deng and Singh 2001; Pannone 2007).

5.2 Evaluation of the Dispersion Coefficient

The dispersion is due to the contact between water particles having different pollutant concentration and increases if the number of particles coming into a mutual contact increases. This can be an effect of the turbulence, which becomes the main factor that activates the dispersion (Magazine et al. 1998; Seo and Maxwell 1992; Kang and Choi 2007).

An accurate investigation requires a precise knowledge of all the geometrical and dynamic characteristics of the water course, and therefore, it is possible only in the case of long pipes under pressure, for which Taylor (1953, 1954) obtained the expression

$$E_L = k\,Ru^* \tag{5.1}$$

where R is the hydraulic radius of the pipe [L], u^* the shear velocity [LT^{-1}] and k a dimensionless constant term. The expression underlines the role of u^* as the most significant term that characterises turbulence. As known from fluid mechanics, the shear velocity in a pipe is

$$u^* = \sqrt{RS_f g} \tag{5.2}$$

where S_f is the slope of the energy line and g the acceleration due to gravity $[LT^{-2}]$. Expressions (5.1) and (5.2) can be adapted to a wide river stretch, where the hydraulic radius can be approximated by the average depth h (Elder 1959):

$$E_L = k\, hu^* \tag{5.3}$$

This assumption is supported by theoretical and experimental investigations, where the effects of the velocity fluctuation have been formally interpreted. Moreover, experiments carried out in laboratory flumes and natural rivers have stressed the importance of the lateral variability of velocity on the river cross section, particularly in wide rivers having irregular bottom and banks.

After recalling some fundamental aspects of the general theory of turbulence, Elder (1959) proposed the value of 5.93 for k in a river. With such an assumption, expression (5.3) is normally used to calculate the longitudinal dispersion coefficient, as the other involved terms can be measured or evaluated with an acceptable reliability.

For an application of the mathematical formulations, it is now essential to have an expression able to give the longitudinal dispersion coefficient as a function of known characteristic terms of the river. A simple way to obtain such a "predictive" expression can be obtained using expression (5.2), where the slope of the energy line, S_f, (dimensionless) can be determined by means of the Manning formula for rivers:

$$U = \frac{1}{n} h^{\frac{2}{3}} j^{\frac{1}{2}} \tag{5.4}$$

with U mean longitudinal velocity component and n roughness coefficient $[L^{-1/3}T]$. The most common values of n, relevant to rivers, have already been presented in Table 4.1, Chap. 4.

After combining expressions (5.2), (5.3) and (5.4), the resulting expression for the coefficient of longitudinal dispersion is

$$E_L = \varepsilon \sqrt{g}\, nU\, h^{\frac{5}{6}} \tag{5.5}$$

The coefficient ε, defined as *dispersion constant*, varies in a very large interval, and there are very few chances to find a correlation with significant characteristic items of the river, such as the flow and the geometry of the cross section, and the presence of bends in the river path (Fisher 1967, 1969). This uncertainty affects the possibility of using expression (5.5) for an acceptable prediction of the dispersion coefficient in a river or stream.

Other formulations have been proposed and confirmed by experimental investigations, with the purpose of facilitating the practical application (Ayyoubzadeh et al. 2004).

Fisher et al. (1979), based on accurate measurements in the laboratory and in several rivers with mean velocity ranging from 0.3 to 1.7 m/s, presented the following simplified equation (in metric units):

$$E_L = 0.011 \frac{U^2 \, w^2}{h u^*} \tag{5.6}$$

where w denotes the river width.

On the other hand, McQuivey and Keefer (1976) proposed the following equation:

$$E_L = 0.058 \frac{Q}{S_f \, w} \tag{5.7}$$

where Q (m^3/s) is the river flow, valid if the Froude number $F = U \, (gh)^{-1/2}$ is smaller than 0.5 (g being the acceleration due to gravity).

More recently, Deng and Singh (2001), Deng et al. (2001, 2002), following experimental investigations in some rivers having different geometry and flow characteristics, proposed another equation for estimating E_L:

$$\frac{E_L}{h u^*} = \frac{0.15}{8 \, \varepsilon} \left(\frac{w}{h}\right)^{\frac{5}{3}} \left(\frac{U}{u^*}\right)^2 \tag{5.8}$$

with

$$\varepsilon = 0.145 + \left(\frac{1}{3,520}\right) \left(\frac{U}{u^*}\right) \left(\frac{w}{h}\right)^{1.38} \tag{5.9}$$

As shown in Table 5.1, the equations presented above give different results that can vary within two orders of magnitude, and this uncertainty an affect the validity of a predictive model.

The examples given in the same table show that there are some ranges where the obtained values of E_L can be more consistent, and this can be a criterion for choosing the most appropriate formula for practical applications. Prudential values are recommended.

The determination of the dispersion coefficient is now a primary task in theoretical and experimental research (Guymer and O'Brien 2000; Guymer 1998; Seo and Cheong 1998; Jirka 2004; Kashefipour et al. 2002; Kashefipour and Falconer 2004; Diamantopoulou et al. 2004; Kim et al. 2007). The problem is approached by means

Table 5.1 Longitudinal dispersion coefficients E_L (m²/s) calculated with the most common equations for some river stretches (m²/s)

	River "A"	River "B"	River "C"	River "D"	River "E"
Fisher et al. (1979)	57.80	15.58	0.87	182.78	577.99
McQuivey and Keeter (1976)	1907.56	4.81	9.63	603.22	190.76
Deng et al. (2002)	48.94	29.26	3.86	154.77	489.44
River main characteristics					
Width (m)	150	15	5	150	150
Depth (m)	3	0.5	0.5	3	3
Slope (–)	0.00001	0.004	0.001	0.0001	0.001
n Manning (m$^{-1/3}$ s)	0.06	0.06	0.06	0.06	0.06
U (m/s)	0.11	0.66	0.33	0.35	1.10
u^* (m/s)	0.02	0.14	0.07	0.05	0.17
Q (m³/s)	49.33	4.98	0.83	156.01	493.34

of unusual procedures, borrowed often from other fields of science, like medicine and biology, among which the neural networks (Piotrowski 2005; Rowinski et al. 2005) and the genetic algorithms seem very efficient (Wallis et al. 2007; Yongcan and Dejun 2007; Tayfur 2009). The dependence of the involved terms on the time (unsteady conditions) is also considered (Guymer and Boxall 2003).

Particular attention is also now devoted on the transverse mixing in the water body (Boxall and Guymer 2003; Yuan et al. 2007) and the secondary flow (Baek and Seo 2005) which, together with the natural turbulence in the stream (Kang and Choi 2007), are the principal cause of dispersion transport in a water body. New perspectives have refreshed the interest also on the longitudinal dispersion coefficient, with the velocity distribution in the cross section being the determinant factor (Chen and Zhu 2007; Deng et al. 2006).

Some studies propose the calculation of the dispersion coefficient by applying the fundamental pollution transport equation in its integrated form to experimental data, in a way that the coefficient becomes the unknown to be determined. Similar approach is due to Haltas (2009), using a stochastic interpretation of transport.

The problem is still open and more refined outcomes are expected, but it is worthy to point out that the results achieved so far are useful to focus on the real dependence on the dynamic aspects of the river, in order to justify the choice of a well-defined value, especially in a predictive application of the model. However, concerning the expected values, it seems that there are no remarkable improvements with respect to the range indicated in Table 5.1.

References

Antonopoulos V, Papamichail D (1991) Stochastic analysis of water quality parameters in stream. In: Tsakiris G (ed) Proceeding of the EWRA conference advances in water resources technology, Athens, Greece. Balkema Publisher, Rotterdam, The Netherlands, pp 369–376

Ayyoubzadeh SA, Faramarz M, Mohammadi K (2004) Estimating longitudinal dispersion coefficient in rivers. In: Proceeding of the 1st Asia-Oceanic Geoscience Society, APHW session, paper 56-RCW-A69, pp 1–8

Baek KO, Seo IW (2005) Estimation of transverse dispersion coefficient based on secondary flow in sinuous channels. In: Proceeding of the 31st IAHR congress, Seoul, Korea (1), p 521 (printed summary)

Bogle G (1997) Stream velocity profiles and longitudinal dispersion. J Hydraul Eng 123(9): 816–820

Boxall JB, Guymer I (2003) Transverse mixing in natural open channels. In: Proceeding of the 30th IAHR congress, Thessaloniki, Greece, theme C, 1, pp 309–315

Carr ML, Rehmann CR (2005) Estimating the dispersion coefficient with an Acoustic Doppler current profiler. World Water and Environmental Resources Congress 2005, Alaska, USA

Chaiwiwatworakul P, Kazama S, Sawamoto M (2005) Influence of hydraulic characteristics to water quality in a river. In: Proceeding of the 31st IAHR congress, Seoul, Korea, vol 1, p 529 (printed summary)

Chen Y, Zhu D (2007) Study on longitudinal dispersion coefficient in trapezoidal cross-section open channels. In: Proceeding of the 32nd IAHR congress, Venezia, Italy, vol 1, p 258 (printed summary)

Czernuszenko W (1987) Dispersion coefficients identification. In: Proceedings of the 22nd IAHR congress, topics on fluvial hydraulics, Lausanne, pp 285–290

Deng ZQ (2002) Theoretical investigation into dispersion in natural rivers. Dissertation, Lund University, Lund, Sweden

Deng ZQ, Jung HS (2007) Scale-dependent dispersion in rivers. In: Proceedings of the 32nd IAHR congress, Venezia, Italy, vol 2, p 571 (printed summary)

Deng ZQ, Singh VP (2001) Longitudinal dispersion in straight rivers. J Hydraul Eng ASCE 127(11):919–927

Deng ZQ, Singh VP, Bengtsson L (2001) Longitudinal dispersion coefficient in straight rivers. J Hydraul Eng 127(11):919–927

Deng ZQ, Bengtsson L, Singh VP, Adrian DD (2002) Longitudinal dispersion coefficient in single-channel streams. J Hydraul Eng 128(10):901–916

Deng ZQ, Bengtson L, Sing VP (2006) Parameter estimation for fractional dispersion model for rivers. Environ Fluid Mech 6:451–475

Diamantopoulou MJ, Antonopoulos VZ, Papamichail DM (2004) The use of a neural network technique for the prediction of water quality parameters of Axios river in Northern Greece. In: Proceedings of the EWRA symposium on water resources management: risk and challenges for the 21st century, Izmir, Turkey, pp 223–234

Elder JW (1959) The dispersion of a marked fluid in turbulent shear flow. J Fluid Mech 5(4): 544–560

Fisher HB (1967) The mechanics of dispersion in natural streams. J Hydraul Div ASCE 93(6): 187–216

Fisher HB (1969) The effect of bends on dispersion in streams. Water Resour Res 5(2):496–506

Fisher HB, List EJ, Koh RCY, Imberger J, Brooks NH (1979) Mixing in inland and coastal waters. Academic, New York, pp 104–138

Fukuoka S, Sayre WW (1973) Longitudinal dispersion in sinuous channels. J Hydraul Div ASCE 99(1):195–217

Givehchi M, Maghrebi M, Abrishami J (2009) New method for determination of depth-averaged velocity for estimation of longitudinal dispersion in natural rivers. J Appl Sci 9(13):2408–2415

Guymer L (1998) Longitudinal dispersion in sinuous channel with changes in shape. J Hydraul Eng ASCE 124(1):33–40

Guymer DR, Boxall J (2003) Assessing transient storage effects in a natural formed meandering channel. In: Proceedings of the 30th IAHR congress, theme B, Thessaloniki, Greece, vol 1, pp 333–348

Guymer L, O'Brien R (2000) Longitudinal dispersion due to surcharged manhole. J Hydraul Eng ASCE 126(2):137–149

Guymer I, West JR (1992) Longitudinal dispersion coefficients in Estuary. J Hydraul Eng ASCE 118(5):718–734

Haltas I (2009) Calculating the macrodispersion coefficient in the stochastic transport equation. In: Proceedings of the 33rd IAHR congress, Vancouver, Canada

Iwasa Y, Aya S (1991) Predicting longitudinal dispersion coefficient in open-channel flow. In: Proceedings of the international symposium on environmental hydrology, Hong Kong

Jirka GH (2004) Mixing and dispersion in rivers. In: Greco M, Carravetta A, Della Morte R (eds) River flow 2004. Taylor & Francis, London

Kang H, Choi S-U (2007) Turbulence modelling of solute transport in open-channel flows. In: Proceedings of the 32nd IAHR congress, Venezia, Italy, vol 2, p 613 (printed summary)

Kashefipour MS, Falconer RE (2004) Longitudinal dispersion coefficients in natural channels. Water Res 36(6):1596–1608

Kashefipour SM, Falconer RA, Lin B (2002) Modelling longitudinal dispersion in natural flows using ANNs. In: Proceedings of river flow 2002, Louvain-la-Neuve, Belgium, pp 111–116

Ki-Chul K, Geon-Hyeong P, Sung-Hee J, Jung-Lyul L, Kyung-Suk S (2011) Analysis on the characteristics of a pollutant dispersion in river environment. Ann Nucl Energy 38:232–237

Kim D, Muste M, Weber L (2007) Software for assessment of longitudinal dispersion coefficients using Acoustic-Doppler Current profiler measurements. In: Proceedings of the 32nd IAHR congress, Venice, Italy, vol 1, p 73 (printed summary)

Liu H (1977) Predicting dispersion coefficient in streams. J Environ Eng Div ASCE 103(1):59–69

Magazine MK, Pathak SK, Pande PK (1998) Effect of bed and side roughness on dispersion in open channels. J Hydraul Eng 114(7):766–782

McQuivey RS, Keefer TN (1976) Convective model of longitudinal dispersion. J Hydraul Div ASCE 102(10):1409–1424

Nordin CF, Troutman BM (1980) Longitudinal dispersion in rivers: the persistence of skewness in observed data. Water Resour Res 16(1):123–128

Pannone M (2007) Longitudinal dispersion in straight channels: the Lagrangian approach. In: Proceedings of the 32nd IAHR congress, Venezia, Italy, vol 2, p 575 (printed summary)

Piotrowski A (2005) Application of neural networks for longitudinal dispersion coefficient assessment. Geophys Res Abstr 7:976

Ranga-Raju KG (1987) Longitudinal dispersion coefficients in open channels. In: Proceedings of the 22nd IAHR congress, topics on fluvial hydraulics, Lausanne, Switzerland, pp 251–257

Riahi-Madvar H, Ayyoubzadeh S (2010) Developing an expert system for predicting pollutant dispersion in natural streams. In: Petrica V (ed) Expert systems. INTECH, Croatia, pp 224–238

Rowinski PM, Piotrowski A, Napiorkowski JJ (2005) Are artificial neural network techniques relevant for the estimation of longitudinal dispersion coefficient in rivers? Hydrol Sci J 50 (1):175–187

Rutherford JC (1994) River mixing. Wiley, Chichester

Seo IW, Baek KO (2004) Estimation of the longitudinal dispersion coefficient using the velocity profile in natural streams. J Hydraul Eng 130(3):227–236

Seo IW, Cheong TS (1998) Prediction of longitudinal dispersion coefficient in natural streams. J Hydraul Eng 124(1):25–32

Seo IW, Maxwell HC (1992) Modeling low-flow mixing through pools and rifles. J Hydraul Eng 118(10):1406–1423

Shen HT (1980) Longitudinal dispersion in natural streams. In: Proceedings of the international conference on water resources development. IAHR, Taipei, China, vol 2, pp 641–650

Singh SK, Beck MB (2003) Dispersion coefficient of streams from tracer experiment data. J Environ Eng 129:539–546

Sooky AA (1969) Longitudinal dispersion in open channels. J Hydraul Div ASCE 95 (4):1327–1346

Tayfur G (2009) GA-optimized model predicts dispersion coefficient in natural channels. Hydrol Res 40(1):65–78

Tayfur G, Singh VP (2005) Predicting longitudinal dispersion coefficient in natural streams by artificial neural networks. J Hydraul Eng 131(11):991–1000

Taylor GI (1953) Dispersion of soluble matter in solvent flowing slowly through a tube. Proc R Soc Lond A 219:186–203

Taylor GI (1954) The dispersion of matter in turbulent flow through a pipe. Proc R Soc Lond A 223:446–468

Wallis SG, Piotrowski A, Rowinski PM, Napiorkowski J (2007) Prediction of dispersion coefficients in a small stream using Artificial Neural Networks. In: Proceedings of the 32nd IAHR congress, Venezia, Italy, vol 2, p 517 (printed summary)

Yongcan Chen, Dejun Z (2007) Study on longitudinal dispersion coefficient in trapezoidal cross-section open channels. In: Proceedings of the 32nd IAHR congress, Venezia, Italy, vol 1, p 258 (printed summary)

Yuan D, Lin B, Tao J, Jian S (2007) Pollutant transport and mixing in water field. In: Proceedings of the 32nd IAHR congress, Venezia, Italy, vol 1, p 252 (printed summary)

Zhen-Gang Ji (2008) Hydrodynamics and water quality. Modeling rivers. Lakes and estuaries. Wiley-Interscience, Hoboken

Chapter 6
The Biochemical Pollution

Abstract The pollutants affecting a surface water body originate from various sources. There is a wide variety of organic pollutants (e.g. pesticides, furans, PAHs, bacteria, viruses, protozoa) usually found in surface water. When decaying, organic pollution is introduced in a surface water body and the free oxygen is depleted in the water. Since aquatic life is suffocated by low oxygen content, the determination of biochemical oxygen demand (BOD) and dissolved oxygen (DO) is a principal indicator for the quality of the surface water body. BOD and DO are often used to estimate the quantity of organic pollutants present in a surface water body, and they deserve a special attention when incorporated in a mathematical model.

6.1 The Most Frequent Type of Pollution

Numerous are the substances transported to a surface water body from its drainage basin, for which the basic concepts described in the previous chapters are applicable. The substances can be classified according to their source, nature and behaviour in contact with water (European Union 2002). In a quite general view, they can be:

- Suspended solids removed from the river bottom and from the banks of the river channel or eroded from the watershed
- Chemical compounds released by weathering processes, as during the hydrological cycle water interacts continuously with the earth
- Chemical compounds produced by anthropogenic activities, either directly introduced in the water or transported in solution, emulsion or in suspension from the ground
- Bacteria and other microorganisms naturally developed in water
- Bacteria, viruses and microorganisms due to human activities
- Gas particles entrapped by water, especially in contact with the atmosphere
- Wet and dry deposition, which includes the flux of all those compounds that are carried to the surface water body by rain and the flux of particles to the surface water body during the absence of rain, respectively

M. Benedini and G. Tsakiris, *Water Quality Modelling for Rivers and Streams*,   57
Water Science and Technology Library 70, DOI 10.1007/978-94-007-5509-3_6,
© Springer Science+Business Media Dordrecht 2013

These substances are normally considered "pollutants" because they alter the original water quality Chapra (1997).

Pollutant interactions in the surface water are subject to continuous changes, being affected by microbiological activity and climatic conditions (particularly precipitation). Additionally, these interactions are controlled by the properties of the earth materials (rocks, soils, sediments), the molecular properties of the pollutants and the geology of each specific location (Berkowitz et al. 2008). A better insight into the effects on the environment and the human life should lead to a better specification, taking into account the fact that not all these substances have a negative impact, but on the contrary, some of them can be beneficial. It is, therefore, more appropriate to call them *water quality indicators*. It must be recalled that, generally speaking, the concept of pollution refers to a quality alteration that makes water not acceptable for various uses. Typical is the example of the domestic wastewater, which cannot be used for human needs, but after some treatment, it can be beneficial for irrigation.

6.2   The BOD

Among the most important quality indicators, the biochemical oxygen demand (BOD) is very significant to characterise the status of a water body. BOD is the first indicator of the pollution status of the river, especially concerning the presence of domestic and urban discharges. Therefore, one of the principal goals in water management practice is the determination of a suitable relationship between the BOD detected in the stream and the pollution sources.

To achieve such a goal, the criterion of *discharging population* is frequently adopted. The resulting pollution load is then calculated after assuming a particular per capita amount of BOD discharged during a reference time. A value of 70 g/day and per capita is generally considered for this purpose, in order to obtain some estimates of the total daily pollution load.

Very often, the municipal sewage conveys also the wastewater coming from industrial activities located in the urban complex. It is, therefore, necessary to evaluate the combined effect of the pollutants discharged in the receiving river. If the industrial wastewater contains pollutants suitable to be characterised by BOD, such an effect can be estimated by the criterion of *equivalent population*. It consists of assuming a number of fictitious inhabitants to which the industrial pollution load can be attributed. The resulting effect can be evaluated in the same way as the effect due to the real population, taking into account the same per capita value of discharged BOD.

There are several procedures suitable for evaluating the equivalent population, and it is always necessary to have an experimental check. The discharged pollution load depends on the type of the productive process and the amount of the industrial output. A procedure frequently adopted assumes that the size of the industrial plant is significantly represented by the number of employees working in the process, which can be transformed in equivalent number of inhabitants by means of a suitable

Table 6.1 Conversion factors of equivalent population for some industrial activities

Industrial activity	χ	Industrial activity	χ
Coal and peat mining	10	Leather and shoe factories	2
Mining of liquid and gaseous fuel	35	Furniture factories	1
Ore mining	40	Joineries	2
Mills and bakeries	1.5	Metallurgical factories	40
Confectioners	205	Metallic carpentry	2
Preserves	17	Construction of electric machinery	1
Dairy farms	37	Processing of non-metallic ore	37
Oil mill	98	Chemical industries	42
Alcohol beverages	205	Coal and petroleum derivatives	40
Tobacco	10	Rubber	37
Textiles (silk, cotton and synthetic fibres)	5	Synthetic textile fibres	40
Wool	5	Paper factories	74
Tailoring and dressmaking	0.6	Printing offices	1

"conversion factor" χ specific for the same process. Table 6.1 gives some values of the conversion factor χ for the most important industrial activities.

The combined effect in the receiving water body is, therefore, evaluated, adding the amount of BOD discharged by the real population to the amount discharged by the equivalent fictitious one. It must be stressed that these considerations are valid only for the industrial wastewater characterised by organic matter behaving in the same way as the wastewater originating from the households (Barbiero and Cicioni 2000; Buraschi et al. 2005; Femia et al. 2005).

In the example illustrated in Fig. 6.1, assuming that the per capita pollution load is 70 g/day of BOD, the urban area of 15,000 inhabitants discharges totally 1,050 kg/day of BOD. In the dairy farm, 20 employees, with a conversion factor $\chi = 37$, correspond to 740 equivalent inhabitants, and the relevant load is 51.8 kg/day of BOD, while in the paper factory, with $\chi = 74$, 100 employees, corresponding to 7,400 equivalent inhabitants, discharge 518 kg/day of BOD. The total pollution load discharged into the river through the sewage collector A-B-D-E is, therefore, 1,618.8 kg/day. If the discharge is entirely diverted to the wastewater treatment plant (path A-B-D-F-G), having an efficiency $\eta = 55\%$, the pollution load is reduced to 728.9 kg/day of BOD.

The criterion of equivalent population cannot be applied to other pollutants coming from the industrial activity, for which direct measurements of the discharged wastewater, in terms of quantity and quality, are necessary.

The above considerations regarding BOD concern the organic matter made up by carbon compounds subject to decay processes characterised by the activity of bacteria, requiring an amount of oxygen properly referred to as *carbonaceous BOD*. This is of paramount importance in the urban and domestic wastewater.

There is also a similar decay process involving matter made up by nitrogen compounds that involves the *nitrogenous BOD* (NBOD), for which analogous considerations can be developed. Normally the NBOD is less important in rivers, and the oxygen required for its decay is considered together with the oxygen

Fig. 6.1 An application of the *equivalent population* method

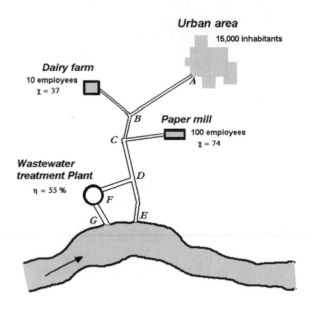

necessary for the more complex processes that characterise the decay of nitrogen compounds, as described in Chap. 7 in more detail.

6.3   The Decay of BOD

The BOD, which includes several chemical and biological reactions somewhat difficult to interpret by means of elementary processes, is a typical nonpersistent water quality indicator. Therefore, as already described in Chap. 4, in the general equation of pollution transport applied to such a quality indicator, the reaction coefficient or *decay term K* is presented as follows:

$$\frac{\partial c}{\partial t} = -K \cdot c \tag{6.1}$$

In Eq. (6.1), the coefficient K plays a fundamental role in determining how the BOD concentration decreases in the water body due to processes that are different from the mechanisms of advection and dispersion. Alternatively, as in preceding chapters, the coefficient K has been defined as the *deoxygenation coefficient*.

The integral of Eq. (6.1), written in ordinary derivatives, from $t = 0$ to t, is

$$c(t) = c(0) \exp(-Kt) \tag{6.2}$$

where $c(0)$, the BOD concentration at time $t = 0$, is the total demand of oxygen necessary to complete the decay of all the organic matter present in water. It is

Fig. 6.2 Typical decay of
BOD in the "bottle"

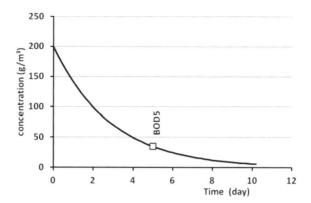

called the *ultimate BOD* (BOD_{ult}), and $c(t)$ is the BOD concentration at any time t.
The development of BOD with regard to time is shown in Fig. 6.2.

The main reason of BOD decreasing is the decomposition and transformation of
organic matter into mineral components, due principally to the action of bacteria,
but a certain number of organic particles present in the water body can also
disappear from the river/stream because they settle on the bottom and enter the
complex mechanisms that characterise the behaviour of sediments.

It is, therefore, more correct to consider K consisting of a *decomposition coefficient*, K_d, and a *settling coefficient*, K_s, in the form

$$K = K_d + K_s \tag{6.3}$$

The settling coefficient is normally referred to in terms of

$$K_s = \frac{v_s}{h} \tag{6.4}$$

in which v_s is the settling velocity of the particle and h the water depth in the river.
As v_s is very small in deep rivers (order of magnitude of a few millimetres per
second), settling is not appreciable and the attention is concentrated on decomposition. Therefore, in practice, one can consider that $K \equiv K_d$.

Equation (6.2) suggests a way of evaluating K in a body of stagnant water having
homogeneous concentration of BOD. Because such a concentration can be frequently measured at different times, K becomes the only unknown. This procedure,
leading to the *bottle BOD*, requires a vessel of small capacity containing an amount
of river water, in which the BOD concentration can be repeatedly measured. The
coefficient K depends also on the water temperature.

Applying the mathematical formulations in a predictive manner, it can be useful
to have for guidance some values related to known characteristic terms of the river.
Researches carried out in the United States (EPA 1987) have found values of K in
the range of 0.02–0.5 day^{-1}, while the decomposition process is affected by the
water depth and is more pronounced in shallow water bodies.

The time dependence of K emphasised by Eq. (6.2) suggests a significant time to which a standard value could be referred to, in order to have comparable measures of BOD. The value relevant to 5 days after the water sample has been collected from the river is generally considered (BOD_5). Expression (6.2) can be used to transform a measurement performed at the generic time t to the BOD_5, adopting a suitable value of K.

The BOD concentration is usually measured in the laboratory with water samples taken from the river; mobile equipments for direct measurements in situ are also available.

For streams and rivers where the pollutant transport takes long time (more than 5 days), the final concentration of BOD normally includes the effect of the nitrogen demand. According to Peirce et al. (1998), the *ultimate BOD* (BOD_{ult}) can be calculated as

$$BOD_{ult} = a(BOD_5) + b(KN) \tag{6.5}$$

where the a and b are constants depending on the specific case, while the term KN (*Kjeldahl nitrogen*) takes into account the nitrogen concentration in both organic and ammonia form.

6.4   The Reaeration Coefficient

Another term to be considered is the reaeration coefficient K_a, already introduced in Chap. 4. In many investigations, it has been given as a function of the stream characteristics. Many attempts have been made in the past to understand its real meaning and to relate it with some other terms more easily measurable (Owens et al. 1964; Dobbins 1964; Gromiec 1989; Tsivoglou and Neal 1976; Tsivoglou and Wallace 1972; Veltri et al. 2007).

One of the most common expressions is due to O'Connor and Dobbins (1956), which in metric units reads

$$K_a = 3.933 \frac{U^{0.5}}{h^{1.5}} \tag{6.6}$$

This equation is valid for water at the temperature of 20 °C and gives the value of K_a in day^{-1}. The term U is the average velocity (m/s) in the river stretch and h the average depth of water in the river (m). Concerning the river, the authors suggest to consider the total liquid volume in the stretch, V, and the area of its free surface, A, assuming

$$h = V/A \tag{6.7}$$

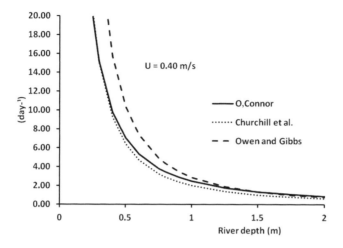

Fig. 6.3 Comparison of different formulations of K_a

Churchill et al. (1962), working on measures taken in streams released from reservoirs where water was undersaturated with oxygen, developed a similar expression

$$K_a = 5.026 \frac{U}{h^{1.67}} \tag{6.8}$$

Based on experiments of artificial reaeration in streams, Owens et al. (1964) proposed

$$K_a = 5.32 \frac{U^{0.67}}{h^{1.85}} \tag{6.9}$$

Expressions (6.6), (6.8) and (6.9) give different values of K_a (days^{-1}) only for the shallower streams, while they give practically the same results in the case of deep rivers, as shown in the example ($U = 0.40$ m/s) described in Fig. 6.3.

Since the formulation of K_a has not been investigated in large streams, the close-to-zero values for the highest depth can be obtained by extrapolation, which is not always justifiable. In a model application aiming at predictive evaluations for large rivers, some prudential values can be recommended. For small rivers, it is advisable to apply the expression developed for a case closer to that described in the preceding paragraphs.

Starting from quite different considerations, Nemerow (1974, 1978) developed an expression that takes into account the real conditions of the stream. In a river stretch of finite length, the author proposed

$$K_a = K \frac{c_u + c_d}{s_u + s_d} - \frac{s_u - s_d}{2.3} \frac{2}{(s_u + s_d)\Delta t} \tag{6.10}$$

where c_u and c_d are, respectively, the BOD concentrations measured at the upstream and downstream cross sections of the river stretch; s_u and s_d are the corresponding DO values; K is the deoxygenation coefficient and Δt is the time needed for the water, with an average velocity U, to pass from one cross section to the other (*travelling time*).

6.5   The Saturation of Dissolved Oxygen

Modelling the DO in a river, particularly in conjunction with BOD or other pollutants consuming oxygen, it is convenient to take the saturation value of dissolved oxygen as the reference term, which is characteristic of the stream, and can be determined with sufficient reliability. Consequently, the real DO situation in the river is evaluated as *deficit from saturation*.

As proved by recent investigations, the concentration S_s of dissolved oxygen saturation depends essentially on water temperature, since other effects (due to salinity and variation of atmospheric pressure with elevation) are less significant and can be neglected. To formulate such dependence, some empirical relationships are available; the most common one is (Commission on Sanitary Engineering 1960a, b)

$$S_s = 14.652 - 0.41022T + 0.0079911T^2 - 0.0000777774T^3 \qquad (6.11)$$

which gives the concentration S_s (expressed in g/m^3) as a function of the temperature T in $°C$.

6.6   Local Oxygenation Sources

In a number of cases, for example, falls, weirs or jumps, as illustrated in Fig. 6.4, the atmospheric air "entrapped" by the water can increase locally the DO concentration in rivers and streams (Nakasone 1987; Bennett and Rathburn 1972; Kim 2005; Kim and Walters 2001; Sousa et al. 2003). To account for these phenomena, several expressions, deduced experimentally, have been proposed.

If the quantity of entrapped air is large, the saturation value can be approximately assumed. Conversely, especially when the drop of the fall is small, the following expression can be used (Avery and Novak 1978):

$$\frac{S_s - s_u}{S_s - s_d} = 10^{0.24\Delta H} \qquad (6.12)$$

Fig. 6.4 Hydraulic
characteristics of a waterfall
as an oxygenation source

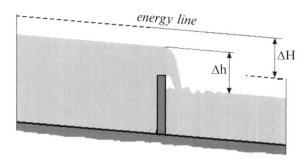

Table 6.2 Coefficient *a*
for the state of pollution

Very polluted	0.65
Moderately polluted	1.00
Slightly polluted	1.60
Clean	1.80

Table 6.3 Coefficient *b*
for type of weir

Flat broad-crested with regular step	0.70
Flat broad-crested with irregular step	0.80
Flat broad-crested with vertical face	0.60
Flat broad-crested with straight-slope face	0.75
Flat broad-crested with curved face	0.45
Round broad-crested with curved face	0.75
Sharp-crested with straight-slope face	1.00
Sharp-crested with vertical face	0.80
Sluice gates	0.05

where s_u and s_d are the values upstream and downstream of the fall, respectively, and ΔH is the total energy loss. ΔH is equal to Δh, the difference of water level from upstream and downstream only in steady uniform flow.

For weirs, a more accurate expression has been worked out by Butts and Evans (1983):

$$\frac{S_s - s_u}{S_s - s_d} = 1 + 0.38ab\Delta h(1 - 0.11\Delta h)(1 + 0.046T) \qquad (6.13)$$

where

Δh = difference of water level (m)
T = water temperature (°C)
a = correction coefficient for the state of pollution in the river
b = correction coefficient for the type of weir

All the terms are in metric units. Values of a and b (dimensionless), for the most common cases, are presented in Tables 6.2 and 6.3, respectively. These considerations concern the local sources of dissolved oxygen, which occur in a single cross section of the river.

Concerning the exchange of oxygen with the atmosphere, particularly important is that occurring on the entire extension of the free surface of the stream. Other possible forms of oxygen injection or subtraction are due to phenomena developing inside the water body, as a result of reactions or mutations involving the pollutants present in water. This matter will be better analysed in Chap. 7. These contributions to the oxygen balance take place in a continuous form along large portion of the water body. Developing a mathematical model and for the sake of simplicity in calculation, it may be convenient to consider a cumulative effect concentrated in an opportune cross section.

References

Avery ST, Novak P (1978) Oxygen transfer at hydraulic structures. J Hydraul Eng 11:1521–1540
Barbiero G, Cicioni G (2000) Report on pollutant loads generated in the pilot basin. Environmental water quality control transnational project, "EWAQC".31 EWAQC-RAP-(2)-T009.0
Bennett P, Rathburn RE (1972) Reaeration in open-channel flow, vol 737, Geological survey professional paper. United States Government Printing Office, Washington, DC
Berkowitz B, Dror I, Yaron B (2008) Contaminant geochemistry. Springer, Berlin/Heidelberg
Buraschi E, Salerno F, Monguzzi C, Barbiero G, Tartari G (2005) Characterization of the Italian lake-types and identification of their reference sites using anthropic pressure factors. J Limnol 64(1):78–84
Butts TA, Evans RL (1983) Effects of channel dams on dissolved oxygen concentrations in Northeastern Illinois Streams, vol 132. State of Illinois, Department of Registration and Education, Illinois State Water Survey, Urbana
Chapra SC (1997) Surface water-quality modeling. WCB-McGraw Hill, Boston
Churchill MA, Elmor HL, Buckingham RA (1962) The prediction of stream reaeration rates. J Sanit Eng Div ASCE 88(SA4):1–46
Commission on Sanitary Engineering (1960a) Solubility of atmospheric oxygen in water. 29th progress report of the commission on sanitary engineering research. J Sanit Eng Div ASCE 86 (SA4):41–53
Commission on Sanitary Engineering (1960b) Effect of water temperature on stream reaeration. 31st progress report of the commission on sanitary engineering research. J Sanit Eng Div ASCE 87(SA6):59–71
Dobbins WE (1964) BOD and oxygen relationship in streams. J Sanit Eng Div ASCE 90 (SA3):53–78
EPA (U.S. Environmental Protection Agency) (1987) The enhanced stream water quality models QUAL2E and QUAL2E-UNCAS. EPA/600/3-87/007. Environmental Research Laboratory, Athens
European Union (2002) Directive 2000/60/EC of the European parliament and of the council of 23 October 2000 establishing a framework for community action in the field of water policy. Official Journal of the European Communities, 22.12.2000 (EN) L327:1–72
Femia A, Barbiero G, Camponeschi S, Greca G, Macri A, Tudini A, Vannozzi M, Vignani D (2005) Economy-wide material flow accounts and derived indicators for Italy: methods and sources. In: OECD-WGEIO workshop on "Material flow indicators and related measurement tools – work session 4: implementation and measurements". Guidance and Best Practice, OECD-WGEIO, Berlin, pp 1–36
Gromiec MJ (1989) Reaeration. In: Jorgensen SE, Groniec MJ (eds) Mathematical submodels in water quality systems. Elsevier, New York, pp 33–64

Kim J (2005) Reaeration test at weirs using oxygen-enhanced environment. In: Proceedings of the 31st IAHR congress, vol 1. Seoul, Korea, p 574 (printed summary)

Kim J, Walters RW (2001) Oxygen transfer at low drop weirs. J Environ Eng 127(7):604–610

Nakasone H (1987) Study of aeration at weirs and cascades. J Environ Eng 113(1):64–81

Nemerow NL (1974) Lectures on water resources management. Italian Water Research Institute, Rome, n. 52–57

Nemerow NL (1978) Industrial water pollution, origins, characteristics, treatment. Addison-Wesley Publishing Co., Reading, UK

O'Connor DJ, Dobbins WE (1956) Mechanism of reaeration in natural streams. J Sanit Eng Div ASCE 82(SA6):1–30

Owens M, Edwards RW, Gibbs JW (1964) Some reaeration studies in streams. Int J Air Water Pollut 8:469–486

Peirce J, Weiner R, Vesilind A (1998) Environmental pollution and control, 4th edn. Butterworth-Heinemann, Boston, An imprint of Elsevier, Amsterdam, Netherland

Sousa C, Lopes R, Matos J, Do Ceu Almeda M (2003) Reaeration by vertical free-fall drops in circular channels. In: A model study, proceedings of the 30th IAHR congress, theme B, Thessaloniki, Greece, pp 361–368

Tsivoglou EC, Neal IA (1976) Tracer measurement of reaeration: III. Predicting the reaeration capacity of inland streams. J Water Pollut Control Fed 48(12):2269–2689

Tsivoglou EC, Wallace JR (1972) Characterization of stream reaeration capacity. Report No. EPA-3-72-012, U.S. Environmental Protection Agency, Washington, DC

Veltri P, Fiorini Morosini A, Maradei G, Verbeni B (2007) Analysis of the re-oxygenation process in water streams. In: Proceedings of the 32nd IAHR congress, vol 1, Venezia, Italy, p 290 (printed summary)

Chapter 7
The Most Frequent Pollutants in a River

Abstract There are numerous pollutants of various nature that affect a water body. They can be chemicals, bacteria or radioactive substances. Many of them are nonconservative and undergo changes when they are in contact with the water. The most frequent pollutants are the compounds of oxygen, nitrogen and phosphorus, whose transformations can be described by means of specific biochemical processes. Their presence in the water quality models is realised in the form of local injection or first-order kinetics.

7.1 Introduction

To better understand the pollutant behaviour and biochemical cycling in the environment, it is important to define the terms "reservoir" and "lifetime or residence time". A reservoir is the place where the components of the biochemical cycle can be retained for long periods of time. Lifetime or residence time is the length of time that a substance is held in a given reservoir. Lifetime is a very useful concept in pollutant cycling. According to Harrison (2007), the residence time is equal to the time taken for the concentration to fall to $1/e$ (where e is the base of natural logarithms) of the initial concentration of the substance, if the source of pollutant is turned off.

7.2 The Oxygen Cycle

Other pollutants, even though considered less significant than BOD, can interact with the dissolved oxygen, as illustrated in Fig. 7.1 (Chang 1998; Millero 2001; James 1979; Prati and Richardson 1971). A better investigation on the DO in water is then opportune. The complex set of reactions affecting DO gives rise to the *oxygen cycle in water* (Chapra 1997; Chevereau 1980).

M. Benedini and G. Tsakiris, *Water Quality Modelling for Rivers and Streams*,   69
Water Science and Technology Library 70, DOI 10.1007/978-94-007-5509-3_7,
© Springer Science+Business Media Dordrecht 2013

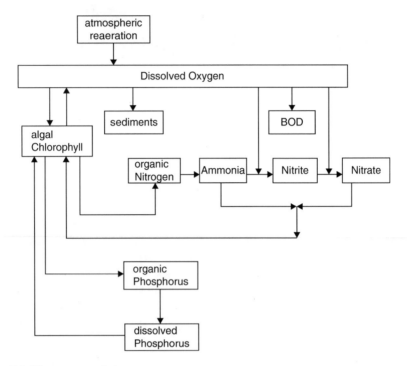

Fig. 7.1 The oxygen cycle in water stream

These pollutants are *nonconservative* (*nonpersistent*), as defined in Chap. 4. Therefore, in the formulation of a mathematical model, they intervene with the time derivative of their concentration, which takes into account not only the variation (decay or growth) in unit time but also the contribution of other phenomena occurring in the water volume, able to increase or decrease the oxygen concentration. The resulting effect is expressed as the sum of the single contributions.

As already mentioned, taking also into account the various reactions presented in Fig. 7.1, the amount of DO in a stream is increased through the direct absorption from the atmosphere, while the bacteria that contribute to the decay of BOD provide a DO subtraction. Moreover, the algae can increase the oxygen content during their growth and subtract it during their death. The sediments, moving vertically along the river depth and settling on the bottom, can also subtract oxygen, as can do some nitrogen compounds, especially ammonia, to be transformed into nitrites and the nitrites in turn to be transformed into nitrates.

The time variation of dissolved oxygen in the river can be expressed as

$$\frac{\mathrm{d}s}{\mathrm{d}t} = K_{\mathrm{a}}(S_{\mathrm{s}} - s) - Kc + D_{\mathrm{A}} - D_{\mathrm{sed}} - D_{\mathrm{amm}} - D_{\mathrm{nitri}} \tag{7.1}$$

where

s = concentration of actual DO $[ML^{-3}]$
S_s = concentration of saturation DO $[ML^{-3}]$
c = concentration of carbonaceous BOD $[ML^{-3}]$
D_A = net variation of DO due to algae activity $[ML^{-3} T^{-1}]$
D_{sed} = uptake of DO due to sediment activity $[ML^{-3} T^{-1}]$
D_{amm} = uptake of DO for ammonia oxidation $[ML^{-3} T^{-1}]$
D_{nitri} = uptake of DO for nitrite oxidation $[ML^{-3} T^{-1}]$

while $K_{a,}$ is the reaeration rate $[T^{-1}]$ and K is the deoxygenation rate of BOD $[T^{-1}]$, which interpret the relevant processes already described in Chaps. 4 and 6.

The DO concentration is evaluated in relation to the saturation value, according to proper chemical considerations (Chapra 1997). Equation (7.1) states also that in case of complete saturation ($s = S_s$), the effect of reaeration is null.

Denoting with σ the *oxygen deficit from saturation*

$$\sigma = S_s - s$$

and remembering that

$$\frac{ds}{dt} = -\frac{d\sigma}{dt}$$

expression (7.1) becomes

$$\frac{d\sigma}{dt} = -K_a\sigma + Kc - D_A + D_{sed} + D_{amm} + D_{nitri} \qquad (7.2)$$

which is the most frequently used form to express the oxygen behaviour in water. It is also worthy to mention that in the majority of cases, the interaction between DO and BOD is predominant and that the last four terms of the Eq. (7.2) can be neglected for simplicity.

The terms of (7.2) refer to the various phenomena that can be observed in the bulk of river water. They require a complex analysis involving chemistry and biology. Their physical meaning and the relevant fundamental mechanisms are summarised in the following paragraphs.

7.3   Algae Activity

A variety of algal species (cyanobacteria, diatoms, greens and macroalgae) can be identified in the river environment. The presence of algae in the river depends on a number of factors including: water quality, temperature, sunlight intensity and

Table 7.1 Frequent values of chlorophyll-*a* concentration in water bodies (mg/m^3)

Type of water body	Min	Max
Highly stirred streams	0.1	1.0
Slow rivers and estuaries	1.0	10.0

presence of chemicals and bacteria (Hornberger and Kelly 1975; Odum 1956; Zhen-Gang 2008).

Since algae uptake dissolves inorganic nutrients during the process of photosynthesis and recycles the nutrients in the forms of inorganic compounds, it is essential for being included in water quality modelling (Zhen-Gang 2008). Moreover, algae affect the biochemical cycles of oxygen, nitrogen and phosphorus primarily through algae death and nutrient uptake. In water quality models, the concentration of algae is typically expressed in biomass as carbon per unit volume. The relationship between the algal biomass as a carbon and algae biovolume can be calculated by the formulas given by Zhen-Gang (2008):

For algae, it reads

$$\log C = -0.460 + 0.866 \log V \tag{7.3}$$

For diatoms only, it reads

$$\log C = -0.422 + 0.758 \log V \tag{7.4}$$

where C = algal biomass in 10^{-12} g as carbon and V = biovolume of algae in 10^{-6} m^3. Since microscopic enumeration of all algae is technically impossible and prohibitively costly, in practice, *chlorophyll-α* represents the total algal biomass because it is much easier to estimate. This becomes the significant indicator of the net algae *productivity*. For a preliminary approach to the problem, some expectable values of chlorophyll-*a* concentration are indicated in Table 7.1.

In rivers and streams, the algae growth is the most important process for algae modelling, while the algal growth rate is a complicated function of nutrients (primarily phosphorus compounds), sunlight, temperature and water turbulence, which is expected to be more significant in the river reaches having the lowest velocity. These conditioning entities are known as *limiting factors*.

The growth of chlorophyll-*a* occurs through the phenomenon of *photosynthesis*, where the light activates some complex reactions with the release of oxygen. The increase of algae concentration is expressed by means of the *growth rate*, μ [T^{-1}], normally variable between 1 and 3 day^{-1}. Its higher values correspond to the higher values of the limiting factors and are expected in a river where the water has been polluted by phosphorus and nitrogen coming from urban or agricultural releases and where the water depth allows a sufficient penetration of sunlight. The process of growth lasts till equilibrium is reached, according to the predominant level of limiting factors.

Table 7.2 Factors κ_p κ_r for the evaluation of DO (mg O_2/mg algae)

Factor	Description	Min	Max
κ_p	Oxygen release	1.2	1.8
κ_r	Oxygen uptake	0.9	2.3

The death of algae occurs primarily through *respiration*, which is the opposite process of photosynthesis. This process, where the dissolved oxygen favours the transformation of the complex molecules into simpler mineral components, is interpreted by means of the *respiration rate*, ρ $[T^{-1}]$, variable from 0.01 to 0.5 day^{-1}. The higher values of such a rate are more appropriate in rivers having high algae concentration.

Both photosynthesis and respiration affect the concentration of DO, and the term D_A in Eq. (7.2), which takes simultaneously into account their effect, can be written as

$$D_A = \alpha_p \mu - \alpha_r \rho \tag{7.5}$$

where α_p is the concentration of oxygen $[ML^{-3}]$ released to the water during the algae photosynthesis and α_r $[ML^{-3}]$ is the concentration of oxygen subtracted from the water during respiration.

Normally α_p is assumed proportional to the total algae concentration, A,

$$\alpha_p = \kappa_p A \tag{7.6}$$

by a factor κ_p representing the amount of oxygen released by the unit concentration of algae. Such a factor is expressed as the ratio of oxygen mass per mass of algae. Concerning the algae concentration, A, the values of Table 7.1 can be assumed.

Similarly for α_r, a factor κ_r is used

$$\alpha_r = \kappa_r A \tag{7.7}$$

which represents the amount of oxygen subtracted by the unit algae concentration.

Some more frequent values of these factors are shown in Table 7.2. The higher values for κ_p are for shallower bodies with low turbidity and medium-high nutrient concentrations. Higher values of κ_r are expected for deeper water bodies.

7.4 Sediment Oxygen Demand

The oxygen demand required for the oxidation of organic matter in benthic sediments is represented by the sediment oxygen demand (SOD). Sediment oxygen demand, BOD and COD are oxygen equivalents; SOD often plays a significant role in affecting dissolved oxygen concentration in water (Zhen-Gang 2008).

The SOD values in a river depend on the sulphide released from the sediment which reacts quickly with the water column when oxygen is available. Furthermore, the sulphide content of the sediment is controlled by the origin of the sediment and the way the sediments are transported in natural waters.

To overcome the many complexities of this process, the behaviour of sediments in a river can be quantified by observing how they settle on the river bottom, which is, therefore, the main reality to take into consideration. For a unit area of the river bottom, the amount of subtracted oxygen is inversely proportional to the river depth, h, in terms of

$$D_s = \frac{\xi}{h} \tag{7.8}$$

being ξ the amount of oxygen requested by the sediments impending on the unit area of river bottom [$ML^{-2} T^{-1}$]. Values of ξ are in the range from 4,000 mg m^{-2} day^{-1} for the sediments originating from urban sewage sludge to 70 mg m^{-2} day^{-1} for the sediments originating from the erosion process of the rocks and soils which dominate the catchment basin.

7.5 Ammonia Oxidation

The oxidation process of ammonia requests a large amount of oxygen in relation to the original quantity of total nitrogen present in the water volume. The process is interpreted by

$$D_{amm} = k_{N,amm}\, \alpha_{amm}\, n_{amm} \tag{7.9}$$

where

$k_{N,amm}$ = rate of transformation of ammonia into nitrite [T^{-1}]
α_{amm} = oxygen uptake by ammonia, expressed as mass of DO per unit mass of ammonia
n_{amm} = concentration of ammonia nitrogen [ML^{-3}]

Values of $k_{N,amm}$ vary between 0.1 and 1.0 day^{-1}. According to Chapra (1997) and Chapra et al. (2006) who conducted experimental investigations, the values of α_{amm} have been found in the range from 3.0 to 4.0 mg of oxygen per mg of nitrogen. The process depends on many factors characterising the stream and on the presence of point pollution sources (sewage discharges). It is expected that streams having high turbulence can favour the process, justifying the higher values of the involved terms.

7.6   Nitrite Nitrogen Oxidation

In the same way, the variation of DO due to the transformation of nitrites into nitrates can be expressed by

$$D_{nitri} = k_{N,nitri}\, \alpha_{nitri}\, n_{nitri} \tag{7.10}$$

where

$k_{N,nitri}$ = rate of transformation of nitrite into nitrate $[T^{-1}]$
α_{nitri} = oxygen uptake by nitrite, expressed as mass of DO per unit mass of nitrogen
n_{nitri} = concentration of nitrite nitrogen $[ML^{-3}]$

Values of $k_{N,nitri}$ range from 0.2 to 2.0 day^{-1}, and values of α_{nitri} vary between 1.0 and 1.14 mg of oxygen per mg of nitrogen (Chapra 1997). Also this process is activated by the presence of bacteria and by the discharge of sewage. Higher values of both $k_{N,nitri}$ and α_{nitri} are expected for shallower and more turbulent water bodies.

7.7   The Nitrogen Cycle

The primary source of all nitrogen species is the elemental nitrogen (N_2) (78% in the atmosphere). The sequential processes of nitrogen compounds transforming organic nitrogen to ammonia, then to nitrite and finally to nitrate, as mentioned in the previous paragraphs, form the *nitrogen cycle*, also important for the evaluation of the water quality in a river (Schnauder and Bockelmann 2005). According to Stumm and Morgan (1996), emission of nitrogen oxides into the atmosphere and nitrogen fertilisation in agriculture have changed the distribution of nitrogen compounds between atmosphere, soil and water.

Every process can be interpreted as function of time, by means of the first derivative of the relevant concentration:

$$\frac{dn_i}{dt} = -k_{N,i} n_i \pm \bar{N} \tag{7.11}$$

where

n_i = concentration of i-th nitrogen compound $[ML^{-3}]$
$k_{N,i}$ = transformation rate of the i-th nitrogen compound $[T^{-1}]$

$\bar{N}$ = nitrogen added or subtracted due to specific phenomena $[ML^{-3}\,T^{-1}]$

The sign of the first term of the right-hand side is minus ($-$) because the transformation acts normally as a decay of the compound.

The specific phenomena, increasing or subtracting the amount of nitrogen, are briefly recalled in the following paragraphs.

Fig. 7.2 Behaviour of organic nitrogen in water

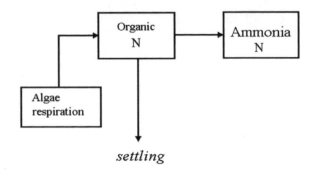

7.7.1 Organic Nitrogen

The organic nitrogen, originally present in water, is partly transformed into ammonia nitrogen and partly settles on the bottom. According to Chapra (1997), the respiration of algae increases the nitrogen content in the water volume, as illustrated in Fig. 7.2.

These transformations can be represented mathematically by

$$\frac{dn_{org}}{dt} = \alpha_A \rho A - k_{N,org} n_{org} - k_{settl} n_{org} \tag{7.12}$$

where

n_{org} = concentration of organic nitrogen $[ML^{-3}]$
A = concentration of algae $[ML^{-3}]$
α_A = fraction of algal biomass expressed in form of nitrogen, quantified as mass of nitrogen per unit mass of algae
ρ = algal respiration rate $[T^{-1}]$
$k_{N,org}$ = rate of transformation of organic nitrogen into ammonia $[T^{-1}]$
k_{settl} = rate of organic nitrogen settling $[T^{-1}]$

The fraction of algal biomass α_A that is nitrogen ranges from 0.07 to 0.09 mg of nitrogen per mg of algae. The most frequent values of $k_{N,org}$ are from 0.02 to 0.4 day^{-1}. Values of k_{settl} vary from 0.001 to 0.1 day^{-1}. The entire process is activated in water bodies having low velocity and high bacteria concentration, conditions that are favourable for choosing the high values of the transformation rate. A conspicuous part of such transformation occurs in the sediment.

7.7.2 Ammonia Nitrogen

As illustrated in Fig. 7.3, the amount of ammonia is increased in the water bulk due to the transformation of organic nitrogen and the amount released by the organisms that are present on the river bottom (the *benthos*).

Fig. 7.3 Transformation of ammonia nitrogen

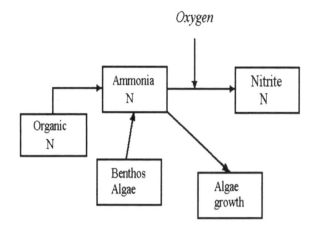

At the same time, the ammonia nitrogen decreases due to its transformation into other compounds as a result of other reactions in the algal biomass. The processes can be approximately represented by

$$\frac{dn_{amm}}{dt} = k_{N,org}\, n_{org} - k_{N,amm}\, n_{amm} \qquad (7.13)$$

where

n_{amm} = concentration of ammonia nitrogen $[ML^{-1}]$
$k_{N,org}$ = transformation rate of organic nitrogen into ammonia $[T^{-1}]$
$k_{N,amm}$ = reaction coefficient for the transformation of ammonia nitrogen into nitrite as an effect of oxidation $[T^{-1}]$

The process is affected by the release of nitrogen from benthos; the subtraction of nitrogen by algae has been found around 0.08 mg of nitrogen per mg of algae. Values of $k_{N,amm}$ range from 0.10 to 1.00 day^{-1}.

7.7.3 Nitrite Nitrogen

The behaviour of nitrite nitrogen in water is relatively simpler, as, according to Fig. 7.4, the amount of nitrite nitrogen increases due to the oxidation of ammonia, but decreases due to its transformation into nitrate:

$$\frac{dn_{nitri}}{dt} = k_{N,amm}\, n_{amm} - k_{N,nitri}\, n_{nitri} \qquad (7.14)$$

where

$n_{,nitri}$ = concentration of nitrite nitrogen $[ML^{-3}]$
$k_{N,nitri}$ = rate of transformation of nitrite nitrogen into nitrate nitrogen due to oxidation $[T^{-1}]$

Fig. 7.4 Transformation of
nitrite and nitrate nitrogen

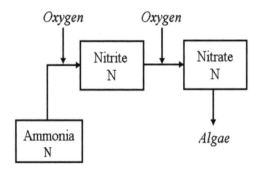

As already specified in Sect. 7.5, the value of $k_{N,nitri}$ varies between 0.2 and 2.0 day^{-1}.

7.7.4 Nitrate Nitrogen

The same Fig. 7.4 illustrates the transformation of nitrite into nitrate, and the final step as well, relevant to the partial subtraction of nitrate in favour of algae. It can be expressed as

$$\frac{dn_{nitra}}{dt} = k_{N,nitri}\, n_2 - \zeta \mu A \qquad (7.15)$$

where

ζ = fraction of algal biomass that is nitrogen, in terms of nitrogen mass per unit mass
   of algae, which is in the range from 0.01 to 0.10 mg of nitrogen per mg of algae

The process is favoured by the presence of other (macro and micro) nutrients. Needless to say, that the presence of nitrogen, in conjunction with other nutrients, is responsible for a conspicuous growth of algae and aquatic plants, which directly impacts on all the aquatic life, subtracting oxygen, and eventually alters the original characteristics of the water body. Such a phenomenon is known as *eutrophication* and has become now a serious threat to river water quality.

7.8   The Phosphorus Cycle

The phosphorus cycle describes the dynamic nature of phosphorus movement within the environment. However, the phosphorus cycle has no major gaseous component; almost all phosphorus in terrestrial ecosystems is derived from weathering of minerals, such as apatite (Merrington et al. 2005). The phosphorus has its own cycle, because in water its organic form, due essentially to the

Fig. 7.5 The phosphorus cycle in water

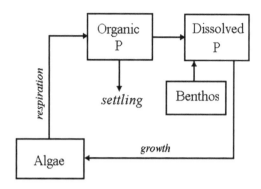

decomposition of dead algae, is transformed into dissolved phosphorus that very often combines with the phosphorus discharged by some external source. The transformation of organic phosphorous, as presented in Fig. 7.5, can be mathematically written as

$$\frac{dp_{org}}{dt} = \pi \rho A - k_{P,org} \, p_{org} - k_{P,settl} \, p_{org} \tag{7.16}$$

where

p_{org} = concentration of organic phosphorus $[ML^{-3}]$
π = phosphorus content in algae, expressed as mass of phosphorus per unit mass of algae
$k_{P,org}$ = rate of organic phosphorus decay $[T^{-1}]$
$k_{P,settl}$ = rate of organic phosphorus settling $[T^{-1}]$

The highest values of π can be 0.02 mg of phosphorus per mg of algae, as observed when a poor sunlight and a low water temperature limit the growth of algae cells.

In turn, the transformations of dissolved phosphorus can be approximately described by

$$\frac{dp_{diss}}{dt} = k_{P,org} \, p_{org} \pm \Pi \tag{7.17}$$

where

p_{diss} = concentration of dissolved phosphorus $[ML^{-3}]$
Π = net amount of dissolved phosphorus due to algae and benthos activity $[ML^{-3} \, T^{-1}]$

while the other terms in the equation have been previously defined. The process depends also on the capability of algae to affect the dissolved phosphorus. The rate coefficient $k_{P,org}$ is in the range between 0.01 and 0.7 day^{-1}, with the higher values occurring during high solar radiation.

In shallow rivers, the release of phosphorus is predominant, and Π can have positive values of the order of 0.01 mg of phosphorus per m^3 of water per day, while in deep rivers, the prevailing phosphorus uptake by the growing algae can give negative values of Π down to -0.6 mg of phosphorus per m^3 of water per day.

7.9   Coliforms

The coliform bacteria, frequently used as indicators of *pathogen contamination*, have also a vital cycle of growth and decay that can be interpreted by means of a first-order expression, normally limited to the die-off phase:

$$\frac{df}{dt} = -k_F f \tag{7.18}$$

where

f = concentration of coliforms, expressed in terms of colonies per unit volume of water

k_F = rate of coliform die-off $[T^{-1}]$

The decay of bacteria is the combined effect of their own natural mortality, the sunlight and the settling on the river bottom. The die-rate of coliforms ranges from 0.05 to 4.0 day^{-1}; the higher values are expected for shallow rivers during high solar radiation.

7.10   The Significant Constants

The coefficients and the constant terms described in the preceding paragraphs are summarised in Table 7.3, with an indication of the most probable values that can be expected in practical cases.

To apply a model in predictive form, the scientific and technical literature can provide description of cases with a large variety of conditions, useful to choose or estimate the proper value of the coefficients mentioned above for the specific case under examination. It is worthy to recommend that this choice should follow an accurate analysis of the river conditions by comparing the adopted values with the observed values.

Table 7.3 Significant terms and their most frequent values

N.	Used symbol	Description	Unit of measure	Remarks (*)	Range Min	Max
1	μ	Algal growth rate	day^{-1}		1	3
2	ρ	Algal respiration rate	day^{-1}		0.05	1.15
3	ξ	Oxygen demand by sediment	$mg\ m^{-2}\ day^{-1}$	t.d.	7.0	4000.0
4	α_{amm}	Oxygen uptake by ammonia	$(mg\ O_2)/(mg\ N)$		3.0	4.0
5	$k_{N,nitri}$	Transformation rate of nitrite into nitrate	day^{-1}	t.d.	0.2	2.0
6	α_{nitri}	Oxygen uptake by nitrite	$(mg\ O_2)/(mg\ N)$		1.0	1.14
7	α_A	Fraction of algal biomass that is nitrogen	$(mg\ N)/(mg\ algae)$		0.07	0.09
8	$k_{N,org}$	Transformation rate of organic nitrogen into ammonia	day^{-1}	t.d.	0.02	0.4
9	$k_{N,amm}$	Transformation rate of ammonia into nitrate	day^{-1}	t.d.	0.10	1.0
10	ζ	Algal biomass that is nitrogen	$(mg\ N)/(mg\ algae)$		0.01	0.09
11	π	Phosphorus contained in algae	$(mg\ P)/(mg\ algae)$		0.01	0.02
12	$k_{P,org}$	Rate of organic phosphorus decay	day^{-1}	t.d.	0.01	0.7
13	$k_{P,settl}$	Rate of organic phosphorus settling	day^{-1}	t.d.	0.001	0.1
14	k_F	Rate of coliform die-off	day^{-1}	t.d.	0.05	4.0
15	k_{settl}	Rate of organic nitrogen settling	day^{-1}	t.d.	0.001	0.1

*t.d. = temperature dependent

7.11 Other Kinds of Pollutants

River water can be polluted by inorganic compounds originating from urban and industrial wastewater. Some inorganic compounds can also be released during the natural contact of water with soils and sediments, like those relevant to sulphur, chromium, arsenic, cadmium and nickel.

The behaviour of these compounds, which are present in the form of metallic ions, as well as the redox reactions and the competitive interactions affecting formation of metal-organic complexes are controlled by the chemical and physical properties of the sediments (Millero and Hershey 1989; O'Brien and Birkner 1977; Pettine and Millero 1990; Stumm and Morgan 1996; Marini et al. 2001; Pham and Waite 2008; Alexakis 2011).

The presence of chemical and biological substances can activate such reactions. Table 7.4 summarises the most common inorganic compounds in river water and their transformations.

Table 7.4 The most common transformations of inorganic compounds in river water

Original substance	Transformed into
Cu (I)	Cu (II)
Fe (II)	Fe (III)
Cr (III)	Cr (VI)
As (III)	As (V)
Mn (II)	Mn (VI)

A redox couple and frequently the dissolved oxygen are responsible for these chemical reactions. Some metallic ions are absorbed by the sediment and by the living organisms that are present in water. Consequently, the overall behaviour of these compounds is the reduction of their concentration, in a way similar to that already described for the organic compounds and bacteria; it can be interpreted by means of the usual first-order kinetic equation:

$$\frac{dc_i}{dt} = \mp k_i c_i \pm C_i \qquad (7.19)$$

where c_i is the concentration of the i-th compound taken into consideration and C_i is a possible external contribution; the rate coefficient k_i is also specific.

Scientific investigations have identified the values of k_i, as clearly described by Millero (2001) and Millero and Hershey (1989), focussing on the relevant chemical equations. At the present time, the results of such investigations have been concentrated on seawater (Pettine and Millero 1990; Benedini et al. 1998), where the original saline content and the stirring of the free surface favour the retention of the oxygen necessary for these transformations, but promising research is also in progress on rivers, where oxygenation process occurs in the upper layer of the stream and in contact with the sediment (Pham and Waite 2008).

Carbon is also an element present in river water, which can act in different ways. Like phosphorus, nitrogen can act as a nutrient, intervening on the algal processes, but its most significant presence is in form of organic compounds, which can affect the oxygen transformations.

Similarly, silicon, originating from the contact with soil-sediment, can be present and undergo several transformations.

In river water, these phenomena are less significant and are quantifiable in an order of magnitude much smaller than that adopted to handle the other phenomena examined in these paragraphs. When developing a general water quality model, particularly for prediction purposes, these considerations enforce the assumption that these pollutants can be considered conservative.

The presence of organic substances is also a reason of particular attention in water quality control, especially the compounds characterised by a complex molecular structure that remain unaltered in contact with water, at least for significant length of time. A full list of organic pollutants is now available, with their origins, the behaviour in water and the analytical methods for their detection and measurement. Table 7.5 summarises the main categories of such principal organic pollutants.

As seen in the table, the organic pollutants are mostly related to special industrial processes; agriculture contributes with the compounds used for agrochemicals, but

Table 7.5 The principal organic pollutants and their most relevant characteristics

Category	Origin	Characteristics
Pesticides	Agriculture	Persistent
Polychlorinated biphenyls (PCB)	Industry	Persistent
Halogenated aliphatics (HAS)	Industry	Persistent
Ethers	Industry	Highly toxic (carcinogenic)
Phthalates	Industry	Persistent
Phenols	Industry	Persistent
Monocyclic aromatics	Industry	Highly toxic
Polycyclic aromatics	Industry	Highly toxic (carcinogenic)
Nitrosamines	Also in current practices	Highly toxic (carcinogenic)

some highly toxic substances can be developed also during normal human daily practices, like the cooking of food.

The behaviour of the organic pollutants in river water is not completely known and depends on many factors, which are not always easy to determine. To treat them in a water quality model, it is advisable to consider them fully conservative, at least as a prudential measure.

7.12 Radioactive Substances

Radioactive pollutants (*radionuclides*) can be released in river water due to the interaction of water with minerals that contain nuclear compounds. Their concentration is generally very low. More significant and harmful can be the presence of nuclear pollutants in the discharge of urban and industrial wastewater.

Radionuclides are decomposed naturally in water, and their behaviour is that of nonpersistent pollutants; a part of their original content is absorbed by the sediment. The radionuclide decomposition takes very long time, of the order of years, following complex reactions that do not fall under the scope of this book. In practice, the nuclear theory suggests to consider the *half-time of decay*. Table 7.6, proposed by Chapra (1997), gives the characteristic values of radionuclides that have recently affected some rivers in North America.

A mathematical model applied to these pollutants requests appropriate time and space scale. In a current application, the radionuclides should be considered persistent.

7.13 Future Perspective and Research Needs

As already mentioned in the preceding sections, the list of pollutants that can reach a river is still open. This largely depends on the discovery of new compounds, following the significant improvement of analytical methods that are able to detect even trace concentrations of substances.

Table 7.6 Characteristic terms of radionuclides of interest for water quality modelling

Radionuclide	Symbol	Half-life (years)	Diffusion coeff. (cm2 s^{-1})	Decay rate (year^{-1})
Plutonium 239,240	^{239,240}Pu	4.5×10^9	12.1×10^{-6}	1.54×10^{-10}
Caesium 137	^{137}Cs	30.0	12.1×10^{-6}	0.023
Strontium 90	^{90}Sr	28.8	3.8×10^{-6}	0.024
Lead 210	^{210}Pb	22.3	4.7×10^{-6}	0.031

Concerning human health, the presence of viruses becomes more and more important. River water having a significant concentration of viruses can become a conveyor of diseases. The behaviour of viruses, organic and inorganic compounds present in natural water requires special investigations, and more information is necessary from the medical sciences, where a thorough knowledge of their characteristics has not yet been achieved.

Concerning the way the viruses can be treated in a water quality model applied to a river, it is reasonable to consider them as nonconservative, with appropriate reaction coefficients to be determined preferably by means of direct infield observations.

Harmful pollutants, as seen in previous paragraphs, are the organic compounds discharged into the river through the sewage or originating from agricultural practices. All these pollutants are related to the human progress in technology, which in turn is dictated by the need of improving the living standard.

It is, therefore, expected in the future to exacerbate the problem of detecting and controlling new pollutants in rivers and streams, stressing the role of mathematical models for water quality assessment.

At the present time, the model structure and algorithms seem to be sufficiently consolidated in expressions and relevant software able to deal with all the pollutants detectable with the actual tools of analysis and measurements. The main difficulty lies in the identification of the appropriate reaction coefficients, specific for each pollutant and able to interpret its behaviour in river water. This rather falls under the scope of aquatic chemistry, geochemistry, biology and toxicology.

References

Alexakis D (2011) Assessment of water quality in the Messolonghi-Etoliko and Neochorio region (West Greece) using hydrochemical and statistical analysis methods. Environ Monit Assess 182:397–413

Benedini M, Passino R, Piacentini G (1998) The quality of receiving bodies in agriculture-dominated areas of Italy. Paper presented at EURAQUA 5th scientific and technical review, Oslo, Norway, pp 93–108

Chang HH (1998) Fluvial processes in river engineering. Wiley, New York

Chapra SC (1997) Surface water-quality modeling. WCB-McGraw Hill, Boston

Chapra SC, Pellettier G, Tao Hua (2006) QUAL2K: a modeling framework for simulating river and stream water quality (version 2.04), U.S. Environmental Protection Agency, Washington DC, USA

Chevereau G (1980) Mathematical model for oxygen balance in rivers. Models for environmental pollution control. Ann Arbor Science Publishers, Ann Arbor, Michigan US: 107–127

Harrison R (2007) Principles of environmental chemistry. RSC Publishing, Cambridge, UK

Hornberger GM, Kelly MG (1975) Atmospheric reaeration in a river using productivity analysis. J Environ Eng Div ASCE 101(EE5):729–739

James A (1979) The value of biological indicators in relation to other parameters of water quality. In: James A, Evison L (eds) Biological indicators of water quality. Wiley, New York, (1): 1–13

Marini L, Canepa M, Cipolli F, Ottonello G, Zuccolini MV (2001) Use of stream sediment chemistry to predict trace element chemistry of groundwater. A case study from the Bisagno valley (Genoa, Italy). J Hydrol 241:194–220

Merrington G, Winder L, Parkinson R, Redman M (2005) Agricultural pollution environmental problems and practical solutions. Taylor & Francis e-Library, Abingdon, UK

Millero FJ (2001) Physical chemistry of natural waters. Wiley Interscience of Geochemistry, London/New York

Millero FJ, Hershey JP (1989) Thermodynamic and kinetics of hydrogen sulfide in natural waters. In: Saltzman ES, Cooper WJ (eds) Biogenic sulphur in the environment, vol 393, ACS symposium. American Chemical Society, Washington, DC, pp 283–313

O'Brien DJ, Birkner FG (1977) Kinetics of oxygenation of reduced sulphur species in aqueous solution. Environ Sci Technol 11:1114–1120

Odum HT (1956) Primary production in flowing waters. Limnol Oceanogr 1(2):102–117

Pettine M, Millero FJ (1990) Chromium speciation in seawater: the probable role of hydrogen peroxide. Limnol Oceanogr 35:730–736

Pham AN, Waite TD (2008) Oxygenation of Fe(II) in natural waters revisited: kinetic modeling approaches, rate constant estimation and importance of various reaction pathways. Geochim Cosmochim Acta 72:3616–3630

Prati L, Richardson QB (1971) Water pollution and self-purification study on the Po River below Ferrara. Water Res 5:203–212

Schnauder I, Bockelmann B (2005) A numerical nitrogen transport model for Carmarthen Bay, Wales, UK. In: Proceedings of the 31st IAHR congress, Seoul, Korea, (2), pp 952–953 (printed summary)

Stumm W, Morgan J (1996) Aquatic chemistry, 3rd edn. Environmental science and technology. A Wiley-Interscience series of texts and monographs. Wiley, New York

Zhen-Gang J (2008) Hydrodynamics and water quality: modelling rivers, lakes and estuaries. Wiley, Hoboken

Chapter 8
Temperature Dependence

Abstract Water temperature represents one of the most significant characteristics of a surface water body. It plays an important role in water quality modelling since it controls many physiological and biochemical processes and affects the solubility of gases and solids.

8.1 Temperature Adjustment

Temperature is one of the most important physical characteristics of surface waters and a crucial factor in water quality modelling. For example, temperature affects the solubility of gases and solids (gases tend to be less soluble in warm water, while solids solubility increases with increasing temperature). Furthermore, water temperature plays an important role on water quality due to a number of reasons:

- Strong influence on some processes such as sorption of organic compounds to particulate matter, volatilisation and reaeration
- Oxygen solubility governed by water temperature (the colder the water, the more the dissolved oxygen)
- Effects on aquatic species that can tolerate only a limited range of temperature
- Control of many physiological and biochemical processes
- Control of the rate of biochemical reactions
- Increasing solubility of solids
- Stratification-destratification and vertical mixing of water
- Arising of eutrophication as the warm water can lower the dissolved oxygen levels

The reaction coefficients described in the previous chapter were specified as *temperature dependent* and referred to the conventional temperature of 20°C. A general expression is proposed to adjust their value (K_i) to the real temperature τ at which they are considered ($K_{i,\tau}$) (EPA 1987):

$$K_{i,\tau} = K_i \vartheta^{\tau-20} \tag{8.1}$$

M. Benedini and G. Tsakiris, *Water Quality Modelling for Rivers and Streams*,
Water Science and Technology Library 70, DOI 10.1007/978-94-007-5509-3_8,
© Springer Science+Business Media Dordrecht 2013

Table 8.1 Factor ϑ for the adjustment of the reaction coefficients as function of temperature

Used symbol	Description	ϑ
K	Deoxygenation rate	1.047
K_a	Reaeration rate	1.024
ξ	Oxygen demand by sediment	1.060
$K_{N,nitri}$	Transformation rate of nitrite into nitrate	1.047
$K_{N,org}$	Transformation rate of organic nitrogen into ammonia	1.047
$K_{N,amm}$	Transformation rate of ammonia into nitrate	1.083
$K_{P,org}$	Rate of organic phosphorus decay	1.047
$K_{P,settl}$	Rate of organic phosphorus settling	1.024
K_F	Rate of coliform die-off	1.047
k_{settl}	Rate of organic nitrogen settling	1.024

The expression depends on an empirical dimensionless factor ϑ, specific for each reaction considered. The most common values of ϑ are shown in Table 8.1.

These considerations underline the importance of water temperature in the evaluation of water quality in a river. It should be recommended that the water temperature accompanies all the measurement to be taken in the water body.

In a river or natural stream, the water temperature depends on various factors and can vary from one instant to the other. The measurement of temperature, therefore, should be performed keeping into account all the aspects in the field of hydrology, hydraulics, aqueous environmental, geochemistry and physics.

8.2   Heat Budget

The total heat budget for a water body includes the effects of heat exchange with the water bottom, the embankments and the atmosphere, the heat generated by chemical and biochemical reactions and the inflow and outflow of water with different temperature (Zhen-Gang 2008).

The atmospheric heat exchange is the major factor which controls the heat budget in nature. More specifically, according to Zhen-Gang (2008), the heat between the water columns and the atmosphere is exchanged in form of:

(a) Turbulent transfer, which includes:

- Sensible heat transfer due to the temperature difference between the overlying air and the water
- Latent transfer due to water evaporation

(b) Radiative process, which includes:

- Long-wave radiation emitted by the water surface and the atmosphere
- Short-wave radiation from the sun

When developing a water quality model, these aspects will be examined with more details in Chap. 16.

References

EPA (U.S. Environmental Protection Agency) (1987) The enhanced stream water quality models QUAL2E and QUAL2E-UNCA (EPA/600/3-87/007). Environmental Research Laboratory, Athens, GA

Zhen-Gang J (2008) Hydrodynamics and water quality: modelling rivers, lakes and estuaries. Wiley, Hoboken

Chapter 9
Application of the General Differential Equations

Abstract With a simplified interpretation of the water body, the fundamental differential equation can be integrated in an analytical way, giving rise to expressions already available in the technical and scientific literature. Such expressions allow for the effect of pollutant injection in a uniform stream to be analysed and are very useful in practice to give a first-glance evaluation of the pollution in the river and streams. Current packages of computing software can be used for an efficient and immediate application.

9.1 An Analytical Solution

The differential equations described in the previous chapters are an interpretation of the pollution transport at the elementary scale, in an infinitesimal volume of the water body. For a more comprehensive view at a larger scale, such equations have to be integrated, with appropriate boundary and initial conditions. Moreover, the knowledge of the dynamic field is necessary, with the appropriate value of water velocity and depth at any point and any instant in the water body, as also explained in the preceding chapters.

The analytical integration of the fundamental differential equations in the most general, three-dimensional and time-dependent case is not easy and entails burdensome mathematical manipulations. The result is a set of complicated formulations, somewhat unwieldy in the current utilisation practice. The new computer facilities, together with the relevant mathematical innovations, are now an efficient tool for the integration in the *numerical field*, after a convenient transformation of the fundamental differential equations into discrete expressions. This will be examined with more details in the following chapters.

In the analytical field, the one-dimensional case is relatively simpler, where all the terms are assumed constant in the y- and z-directions and depend only on the

M. Benedini and G. Tsakiris, *Water Quality Modelling for Rivers and Streams*,
Water Science and Technology Library 70, DOI 10.1007/978-94-007-5509-3_9,
© Springer Science+Business Media Dordrecht 2013

Fig. 9.1 Schematic
interpretation of the pollutant
injection in one-dimensional
stream

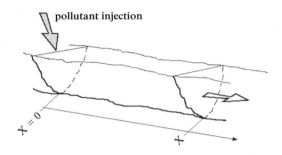

longitudinal coordinate x. In practice, this is an approach to the behaviour of a stream or a long river stretch, where the quality characteristics can be considered constant at any cross section, but variable along stretches into which the water body can be split. The constant values in a cross section normally come from integration along the vertical direction z and assuming an average value along the transverse direction y.

The one-dimensional case of nonconservative pollutant without external contributions, which will be considered in the following paragraphs, is also useful to understand some particular aspects of the complex phenomena of pollution transport.

Under these assumptions, the fundamental differential Eq. (4.8) of Chap. 4 becomes

$$\frac{\partial C}{\partial t} + \frac{\partial v_x C}{\partial x} = E\frac{\partial^2 C}{\partial x^2} - kC \tag{9.1}$$

It can be integrated following a sequence of steps described in many mathematical textbooks and already applied to particular problems of water quality (O'Loughlin and Bowmer 1975; Runkel 1996).

It is assumed that both water velocity v_x and the cross-sectional area A_x are known at any value of x and t.

By means of this differential equation, several problems of pollution transport in a water body can be approached and solved. The simplest problem concerns the evaluation of the pollutant concentration along the stream. As sketched in Fig. 9.1, if the pollutant concentration is known at the cross section $x = 0$, it can be evaluated at any time at any downstream cross section.

According to the one-dimensional assumption, there is no variation of the pollutant concentration on the cross-sectional area and, in an acceptable order of approximation, this allows also to interpret the case in which the abscissa $x = 0$ identifies the place in which the pollutant is injected into the stream.

The problem consists of finding the concentration $C = C(x,t)$ in a downstream cross section located at $x > 0$ and at the instant $t > 0$.

Two significant cases will be considered in a rectilinear channel having uniform conditions ($v_x = constant = U$ and $A_x = constant$).

9.1.1 Continuous Source of Infinite Duration

The pollutant concentration is expressed in dimensionless form as a percentage of the original value C_0.

At cross section $x = 0$, the pollutant is injected continuously, and the concentration C_0 remains indefinitely constant.

The stream is assumed to be undisturbed $(C = 0)$ for its entire length, at the beginning $(t = 0)$, as well as before the presence of pollutant can be appreciated. This is interpreted by the initial and boundary conditions

$$C(x,0) = 0.0 \quad \text{for} \quad x \geq 0$$
$$C(0,t) = C_0 \quad \text{for} \quad t \geq 0$$
$$C(\alpha,t) = 0.0 \quad \text{for} \quad t > 0$$

Integration of (9.1) gives the equation

$$C(x,t) = \frac{C_0}{2}\left\{\exp\left[\frac{Ux}{2E}(1-\Gamma)\right]\text{erfc}\left(\frac{x-Ut\Gamma}{2\sqrt{Et}}\right) + \exp\left[\frac{Ux}{2E}(1+\Gamma)\right]\text{erfc}\left(\frac{x+Ut\Gamma}{2\sqrt{Et}}\right)\right\}$$

$$(9.2)$$

where

$$\Gamma = \sqrt{1 + 4\frac{kE}{U^2}}$$

This expression is valid only for $t > 0$ and for the presence of all the various phenomena examined. The dispersion mechanism cannot be ignored $(E > 0)$. The expression can be applied to some specific cases and is useful to understand the role and importance of the various mechanisms involved in pollutant transport. The expression can be easily worked out using one of the common spreadsheets available in the computer software. As an alternative, an ad hoc code can be developed using a common computer language.

An application is described in Fig. 9.2, where the time-dependent variation of $C = C(x,t)$ is shown for significant cross sections, located, respectively, at $x = 100.0$ m and $x = 1000.0$ m. It is easy to observe that, after a transition interval, the concentration reaches the "saturation" value, close to C_0, which remains constant for the rest of the time. The length of transition interval depends on the location of the cross section considered downstream from the injection: the more downstream is this cross section, the longer is the time interval during which the achievement of the saturation is delayed.

The same Eq. (9.1) is useful to explain the effect of the reaction coefficient k, as shown in Fig. 9.3.

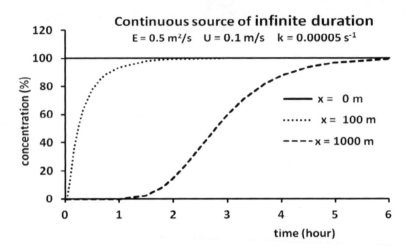

Fig. 9.2 Pollutant concentration in a rectilinear uniform channel as function of time, at various cross sections downstream from the injection

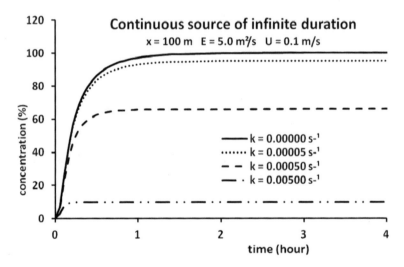

Fig. 9.3 Abatement of nonconservative pollutants characterised by various reaction coefficients

Four situations are examined. The first is for a conservative pollutant ($k = 0.00\,\mathrm{s}^{-1}$); the others concern increased values of reaction coefficient. The pollutants having the highest values will never reach saturation, as the abatement due to reaction predominates over the other transport phenomena.

The effects of dispersion are examined in Fig. 9.4, after applying Eq. (9.2) with different values of the coefficient E.

When stressing the role of dispersion by the assumption of high values of E, the pollution transport is activated since the first instants of the process, but it takes

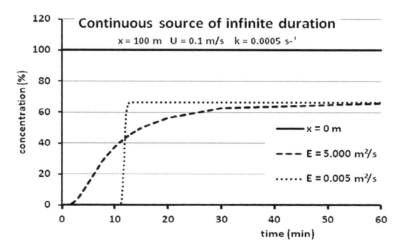

Fig. 9.4 Effect of dispersion on the pollutant behaviour

more time to reach a constant value through a continuous progressive mixing. Vice versa, in the case of predominant advection, with very low values of E, the effect of transport can be appreciated only after the time correspondent to the flow velocity has elapsed, but the constant value is achieved in a very short time.

As already shown in Fig. 9.3, the relatively high reaction coefficient does not allow for saturation to be reached.

9.1.2 Source of Finite Duration

The case of a constant injection of finite duration can be examined by means of the same Eq. (9.1). Also in this case, the stream is assumed undisturbed, for its entire length, at the beginning and before the pollutant presence, according to the boundary conditions

$$C(x,0) = 0.0 \quad \text{for} \quad x = 0$$
$$C(0,t) = C_0 \quad \text{for} \quad 0 < t \leq \tau$$
$$C(0,t) = 0.0 \quad \text{for} \quad t > \tau$$
$$C(\infty,t) = 0.0 \quad \text{for} \quad t \geq 0$$

where τ is the duration of pollutant injection.

The resulting equation refers to two different time intervals, respectively, for $t \leq \tau$ and $t > \tau$, according to the *principle of superposition* ("the resulting effect is the sum of the single effects"). During the first time interval, the effect of injection

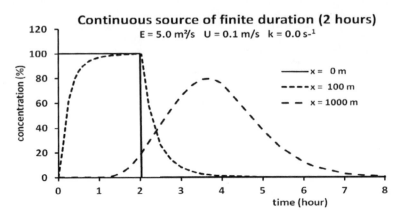

Fig. 9.5 Pollutant transport as function of time, at the injection and various downstream cross sections

is still interpreted by Eq. (9.2), while after the injection has stopped, the effect is interpreted by subtracting from the same equation (supposed to last indefinitely) what should be the effect beginning at $t = \tau$. In other words, the case is represented by the two equations:

$$C(x,t) = \frac{C_0}{2}\left\{\exp\left[\frac{Ux}{2E}(1-\Gamma)\right]\mathrm{erfc}\left(\frac{x-Ut\Gamma}{2\sqrt{Et}}\right) + \exp\left[\frac{Ux}{2E}(1+\Gamma)\right]\mathrm{erfc}\left(\frac{x+Ut\Gamma}{2\sqrt{Et}}\right)\right\}$$

(9.3a)

for $0 < t \le \tau$
and

$$C(x,t) = \frac{C_0}{2}\left\{\exp\left[\frac{Ux}{2E}(1-\Gamma)\right]\left[\mathrm{erfc}\left(\frac{x-Ut\Gamma}{2\sqrt{Et}}\right) - \mathrm{erfc}\left(\frac{x-U(t-\tau)\Gamma}{2\sqrt{E(t-\tau)}}\right)\right]\right.$$
$$\left. + \exp\left[\frac{Ux}{2E}(1+\Gamma)\right]\left[\mathrm{erfc}\left(\frac{x+Ut\Gamma}{2\sqrt{Et}}\right) - \mathrm{erfc}\left(\frac{x+U(t-\tau)\Gamma}{2\sqrt{E(t-\tau)}}\right)\right]\right\}$$

for $t > \tau$.

An application of these equations, for a high dispersion pollutant, is shown is in Fig. 9.5, where the presence of pollutant appears delayed, depending on the values of the involved terms. In such a situation, the pollutant presence in the downstream cross sections is attenuated.

Finally, the effect of velocity, responsible of advection, is examined in Fig. 9.6.

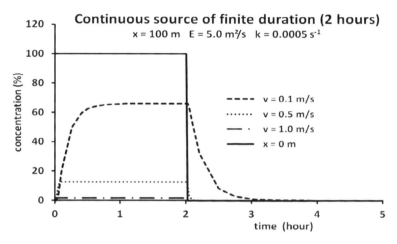

Fig. 9.6 The effect of velocity on pollutant transport

The higher the velocity is, the shorter is the time interval necessary to achieve a constant value of the concentration, which, in turn, decreases as the pollutant is easily removed from the water.

The examples and the particular cases described are useful to show the effects relevant to the significant terms inserted in the mathematical models, which are determinant for the correct understanding of the water quality in the water body under examination.

9.2 Some Comments

The preceding paragraphs have shown some peculiarities of the analytical solution of the fundamental differential equation of pollutant transport. This is due to the very simple way of interpreting the reality, by means of a one-dimensional approach. All the significant terms of the problem are considered, with the only exception of the external contribution, the insertion of which could bring in some formal complexity.

Neglecting the external contribution can be a limitation in case of a river receiving several concentrated pollutant injections. A possible way of overcoming such an impasse is to split the river into a sequence of segments, each one having its origin in the cross section in which the injection occurs, and to apply the expression with its initial point in that section. The initial concentration C_0 will result from the amount of pollutant injected in the same section plus the residual of the quantity injected at the beginning of the upstream segment, after its abatement along the river stretch.

Concerning the other fundamental terms, it is worthy to point out how they intervene in the overall process of pollution abatement. In particular, the k factor underlines the effect of "self-purification", normally accepted in a river where there are a limited number of pollutant discharges. The expressions described in this section have been the object of interesting applications for a first-glance characterising of the water quality in a river.

The identification of a particular case of finite duration injection can provide a tool able to interpret complex hydraulic and quality situations (Franco et al. 2000). The opportunity of applying the described concepts has therefore increased.

9.3   Computing Procedures

Many advanced software packages are now available in the computer market, able to interpret and implement the proposed expressions, with the possibility of analysing real situations characterised by numerous variables. The focus of the problem still remains on the validity of the analytical approach, because it is always uncertain how a river, having complex geometrical and hydraulic connotations, can be represented by a one-dimensional and a simplified tool. In spite of the examples mentioned in the preceding paragraphs, transferring the analytical approach to a real water body obliges always to consider burdensome mathematical adjustments, without the certainty to have completely understood the reality of the problem.

The analytical approach can be, therefore, useful only for an immediate and generic evaluation, leaving to other more sophisticated tools the possibility to give a more accurate insight of the problem. In this concern, the application of the expressions described in the preceding paragraphs can benefit from simplified computing procedures that are also easily available in the computer environment.

Among them, the spreadsheet proposed by some software firm (principally "Microsoft Excel" and "Lotus 123") can be very useful. They contain, already codified, the necessary mathematical formulations and special routines for the graphical presentation of the final results. Their application does not require special knowledge in the computer science and can be done very easily by the persons responsible for water problems, becoming in this way a powerful tool to enhance their own professional expertise.

The application of these procedures entails the adoption of sequences of discrete values of the variables and all the other terms taken into consideration, but there is no restriction on how such values can be taken (unlike in the case of the numerical procedures that will be described in the following chapters), giving the operator plenty of freedom to choose the values that are more significant for the correct approach to the problem.

	A	B	C	D	E	F	G	H	I	J	K	L	M	N	O	P	Q	R
1																		
2																		
3	a	b	c	d	e	f	g	h	i	j	k	l	m	n	o	p	q	r
4	0.1	1.00	0.00	1.00	0.01	99.99	1.41	70.70	0.00	2.00	7.39	100.01	70.72	0.00	0.00	0.00	0.00	0.00
5	0.2	1.00	0.00	1.00	0.02	99.98	2.00	49.99	0.00	2.00	7.39	100.02	50.01	0.00	0.00	0.00	0.00	0.00
6	0.3	1.00	0.00	1.00	0.03	99.97	2.45	40.81	0.00	2.00	7.39	100.03	40.84	0.00	0.00	0.00	0.00	0.00
7	0.4	1.00	0.00	1.00	0.04	99.96	2.83	35.34	0.00	2.00	7.39	100.04	35.37	0.00	0.00	0.00	0.00	0.00
8	0.5	1.00	0.00	1.00	0.05	99.95	3.16	31.61	0.00	2.00	7.39	100.05	31.64	0.00	0.00	0.00	0.00	0.00
9	1	1.00	0.00	1.00	0.10	99.90	4.47	22.34	0.00	2.00	7.39	100.10	22.38	0.00	0.00	0.00	0.00	0.00
10	2	1.00	0.00	1.00	0.20	99.80	6.32	15.78	0.00	2.00	7.39	100.20	15.84	0.00	0.00	0.00	0.00	0.00
11	1800	1.00	0.00	1.00	180.00	-80.00	189.74	-0.42	1.45	2.00	7.39	280.00	1.48	0.04	1.45	0.27	1.72	86.08
12	3600	1.00	0.00	1.00	360.00	-260.00	268.33	-0.97	1.83	2.00	7.39	460.00	1.71	0.02	1.83	0.11	1.94	97.14
13	5400	1.00	0.00	1.00	540.00	-440.00	328.63	-1.34	1.94	2.00	7.39	640.00	1.95	0.01	1.94	0.04	1.99	99.26
14	7200	1.00	0.00	1.00	720.00	-620.00	379.47	-1.63	1.98	2.00	7.39	820.00	2.16	0.00	1.98	0.02	2.00	99.79
15	10800	1.00	0.00	1.00	1080.00	-980.00	464.76	-2.11	2.00	2.00	7.39	1180.00	2.54	0.00	2.00	0.00	2.00	99.98
16	14400	1.00	0.00	1.00	1440.00	-1340.00	536.66	-2.50	2.00	2.00	7.39	1540.00	2.87	0.00	2.00	0.00	2.00	100.00
17	18000	1.00	0.00	1.00	1800.00	-1700.00	600.00	-2.83	2.00	2.00	7.39	1900.00	3.17	0.00	2.00	0.00	2.00	100.00
18	21600	1.00	0.00	1.00	2160.00	-2060.00	657.27	-3.13	2.00	2.00	7.39	2260.00	3.44	0.00	2.00	0.00	2.00	100.00
19	---	---	---	---	---	---	---	---	---	---	---	---	---	---	---	---	---	---
20	---	---	---	---	---	---	---	---	---	---	---	---	---	---	---	---	---	---

Fig. 9.7 Image of the spreadsheet used for the effect of a continuous source of infinite duration of a persistent pollutant. The content of the columns is specified in Table 9.1

The examples described in this chapter have been developed by means of a common spreadsheet, and Fig. 9.7, with the support of Table 9.1, can give a hint on how to proceed for analysing a simple problem.

The problem data are

$$C_0 = 100\,\% \quad U = 0.10 \text{ m/s}$$
$$x = 100 \text{ m} \quad E = 5.00 \text{ m}^2/\text{s}$$
$$K = 0.00 \text{ s}^{-1} \quad \Gamma = \sqrt{1 + 4\frac{KE}{U^2}} = 1.00$$

to which the values in the various columns are referred, according to the functions and the specification of Table 9.1.

The final result is the concentration $C(t)$ at the cross section $x = 100.00$ m from the injection.

Table 9.1 Content of the columns of the spreadsheet shown in Fig. 9.7

Column	Expression	Comment	Column	Expression	Comment
A	Time (t)	Variable	J	$\dfrac{Ux}{2E}(1+\Gamma)$	
B	$\dfrac{Ux}{2E}$		K	$\exp\left[\dfrac{Ux}{2E}(1+\Gamma)\right]$	
C	$\left[\dfrac{Ux}{2E}(1-\Gamma)\right]$		L	$x+Ut\Gamma$	
D	$\exp\left[\dfrac{Ux}{2E}(1-\Gamma)\right]$		M	$\dfrac{x+Ut\Gamma}{2\sqrt{Et}}$	
E	$Ut\Gamma$		N	$\mathrm{erfc}\left(\dfrac{x+Ut\Gamma}{2\sqrt{Et}}\right)$	
F	$x-Ut\Gamma$		O	$\exp\left[\dfrac{Ux}{2E}(1-\Gamma)\right]\mathrm{erfc}\left(\dfrac{x-Ut\Gamma}{2\sqrt{Et}}\right)$	
G	$2\sqrt{Et}$		P	$\exp\left[\dfrac{Ux}{2E}(1+\Gamma)\right]\mathrm{erfc}\left(\dfrac{x\pm Ut\Gamma}{2\sqrt{Et}}\right)$	
H	$\left(\dfrac{x-Ut\Gamma}{2\sqrt{Et}}\right)$		Q	$\exp\left[\dfrac{Ux}{2E}(1-\Gamma)\right]\mathrm{erfc}\left(\dfrac{x-Ut\Gamma}{2\sqrt{Et}}\right)+\exp\left[\dfrac{Ux}{2E}(1+\Gamma)\right]\mathrm{erfc}\left(\dfrac{x+Ut\Gamma}{2\sqrt{Et}}\right)$	
I	$\mathrm{erfc}\left(\dfrac{x-Ut\Gamma}{2\sqrt{Et}}\right)$		R	$\dfrac{c_0}{2}\left\{\exp\left[\dfrac{Ux}{2E}(1-\Gamma)\right]\mathrm{erfc}\left(\dfrac{x-Ut\Gamma}{2\sqrt{Et}}\right)+\exp\left[\dfrac{Ux}{2E}(1+\Gamma)\right]\mathrm{erfc}\left(\dfrac{x+Ut\Gamma}{2\sqrt{Et}}\right)\right\}$	Final result

References

Franco C, Paoletti A, Sanfilippo U (2000) A new approach for modelling organic solute transport into rivers: the Instantaneous Unit Pollutograph (IUP) mode. In: Proceedings of the international conference hydroinformatics, Iowa City, IA, USA

O'Loughlin EM, Bowmer KH (1975) Dilution and decay of aquatic herbicides in flowing channels. J Hydrol 26:217–235

Runkel R (1996) Solution of the advection-dispersion equation: continuous load of finite duration. J Environ Eng 122(9):830–832

Chapter 10
The Steady-State Case

Abstract In the one-dimensional approach, the case of pollutant injection and river behaviour that do not depend on time is frequently considered. The integration of the fundamental differential equation is simplified and gives rise to very useful expressions for an overall assessment of water pollution. A further simplification is the case of nondispersive flow, which, applied to the BOD and DO interaction, results in the well-known Streeter and Phelps model, currently used for a first-glance evaluation of a river contamination caused by urban wastewater discharge.

10.1 The Fundamental Equation in One-Dimensional Approach

The preceding chapter has dealt with the pollution transport in its most general expression, in which the pollutant concentration, also in its simplified one-dimensional form, is a function of both time and space, $C = C(x,t)$. The cases considered as an example are based on the assumption that the pollutant injection in the stream starts at the instant $t = 0$ at the cross section located at $x = 0$. This implies that for $t < 0$ and $x < 0$, no injection occurs. As seen in the preceding chapter, such an assumption gives rise to a transition time, infinite under the mathematical viewpoint, but in practice very long, and the stable conditions can be expected only at the end of the transition time. The assumption allows also to consider a pollutant injection variable with time, like in the event of accidentally pouring a contaminant substance into the river.

In the current practice, the steady state, where the pollutant injection does not depend on time, is often considered. The pollution transport is examined, therefore, along the river stretch in relation to a constant injection assumed to last continuously, like the discharge of an urban sewage.

Under this assumption, the general differential introduced in Chap. 4 becomes

$$0 = -\left(\frac{\partial v_x C}{\partial x} + \frac{\partial v_y C}{\partial y} + \frac{\partial v_z C}{\partial z}\right) + E\left(\frac{\partial^2 C}{\partial x^2} + \frac{\partial^2 C}{\partial y^2} + \frac{\partial^2 C}{\partial z^2}\right) \pm S - kC \quad (10.1)$$

M. Benedini and G. Tsakiris, *Water Quality Modelling for Rivers and Streams*,
Water Science and Technology Library 70, DOI 10.1007/978-94-007-5509-3_10,
© Springer Science+Business Media Dordrecht 2013

in which it is assumed permanently $C = C(x,y,z)$.

Also, in this case, the analytical integration is not easy, requesting several mathematical manipulations and giving rise to complicated formulations. If the one-dimensional approach is followed, several simplifications can be introduced.

In the one-dimensional case, being $C = C(x)$ function of only one variable, neglecting also the external contributions, Eq. (10.1) is transformed into an ordinary differential equation

$$0 = E\frac{d^2C}{dx^2} - U\frac{dC}{dx} - kC \tag{10.2}$$

with U the average velocity in the x-direction, assumed constant all over the stretch considered in the river.

The integral of (10.2), for $x > 0$, is

$$C = C_0 \exp(jx) \tag{10.3}$$

where

$$j = \frac{U}{2E}\left(1 - \sqrt{1 + \frac{4kE}{U^2}}\right)$$

with C_0 the concentration of the injected pollutant.

10.2   The Nondispersive Flow

The dispersion process in pollution transport is very frequently ignored in rivers, and the ordinary differential Eq. (10.2), for $E = 0$, becomes

$$0 = -U\frac{dC}{dx} - kC \tag{10.4}$$

the integral of which is

$$C = C_0 \exp\left(-\frac{kx}{U}\right) \tag{10.5}$$

According to such an approach, the pollutant transport, at various time instants, occurs in a sequence of uniform "channels", as sketched in Fig. 10.1, which replaces a river stretch of known length. In each channel, the pollutant concentration, extended on the whole river cross section, remains constant.

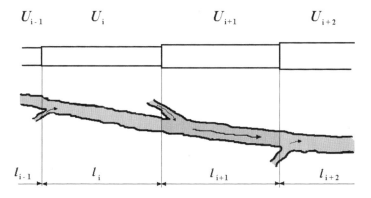

Fig. 10.1 Interpretation of a river with the plug flow approach

From one channel to the other, the pollutant concentration varies only as a function of the decay term. This configuration refers to the *plug flow* approach, very frequently adopted in practice.

A very common case is the evaluation of the BOD/DO interaction, where the differential Eq. (10.4) is applied for the two pollution indicators, as

$$0 = -U\frac{dC}{dx} - KC \qquad (10.6a)$$

for BOD, C being its concentration and K the deoxygenation coefficient, and

$$0 = -U\frac{ds}{dx} + KC - K_a s \qquad (10.6b)$$

for the oxygen deficit $\sigma = S_s - s$, being S_s the saturation value and s the actual concentration of dissolved oxygen, with K_α the reaeration coefficient. These equations must be considered jointly, and, following ordinary mathematical procedures, their integrals are, respectively,

$$C = C_0 \exp\left(-\frac{Kx}{U}\right) \qquad (10.7a)$$

$$\sigma = \sigma_0 \exp\left(-\frac{K_a x}{U}\right) + \frac{KC_0}{K_a - K}\left[\exp\left(-\frac{Kx}{U}\right) - \exp\left(-\frac{K_a x}{U}\right)\right] \qquad (10.7b)$$

in which C_0 and σ_0 are the concentrations of BOD and DO at $x = 0$.

Equations (10.7a) and (10.7b) give the concentration of BOD and DO along the river stretch. They are known in the scientific literature as the *Streeter and*

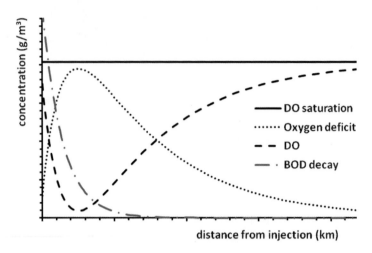

Fig. 10.2 Qualitative response of the Streeter and Phelps model

Phelps model, which for many years has interpreted the behaviour of a river subject to the discharge of urban pollutants (Streeter and Phelps 1925).

Several modifications have been introduced in the Streeter and Phelps model, including also the possibility of incorporating the effect of BOD settling, that implies the adoption of a new reaction coefficient to be added to the deoxygenation coefficient K (Chapra 1997). An attempt has been made for analysing the behaviour of a stream completely devoid of oxygen, in *anaerobic condition,* a case which frequently occurs in a river when the discharge of urban sewage exceeds the natural capability of BOD decay.

Equations (10.7a) and (10.7b) elucidate the importance of the variability of BOD and DO concentrations as function of time. In a river reach between two cross sections at x_1 and x_2, respectively, where the water velocity is U, the expression

$$T = \frac{x_2 - x_1}{U} \tag{10.8}$$

gives the *travelling time T* between the two cross sections.

Following expression (10.8), the Streeter and Phelps model can be rearranged as function of the travelling time in order to analyse how the concentration varies from one section to the other.

A graphical interpretation of the Streeter and Phelps model is shown in qualitative form in Fig. 10.2.

The concentration variability takes place in long river stretches, of the order of several kilometres, depending on the value of the reaction coefficients. It is worthy to notice the DO behaviour, which initially decreases then slowly increases, tending to the saturation value. Such a characteristic behaviour, experimentally confirmed, is known as the *oxygen sag.*

A steady-state model is appropriate for a limited group of water bodies, such as rivers with low variability of pollutant loads on an annual scale and low variability of flow. Besides the peculiarities of the model, it is worthy to notice that, since it neglects the dispersion pollutant transport, it cannot interpret completely the real behaviour of a river, especially when a low water velocity emphasises the role of dispersion. In this case, the model can give only a qualitative indication of the natural phenomena, and more refined tools are necessary also to fit the experimental data that can be collected in the field.

Attempts have been made to integrate Eqs. (10.6a) and (10.6b) after including the dispersion term, but the resulting expression is not always easy to apply in the majority of concrete cases.

For a more reliable interpretation of the pollutant behaviour in a river, the numerical procedures, which are explained in the following chapters, can achieve a more realistic representation of the processes involved.

References

Chapra SC (1997) Surface water-quality modeling. WCB-McGraw Hill, Boston
Streeter HW, Phelps EB (1925) A study of the pollution and natural purification of the Ohio River. US Public Health Service. Public Health Bull 146:75

Chapter 11
Interpretation in Finite Terms

Abstract The real complexity of river geometry and the process of pollutant injection cannot always be modelled accurately by means of the analytical approach. The progress in numerical modelling allows for a rational interpretation of finite terms, supported by the increased computational capacity. The reality can be, therefore, discretised, so that the details of the problem can be better analysed. Therefore, the fundamental differential equation can be transformed into an expression in finite terms for simulating the behaviour of the river. There are several forms of this approach; however, a simple procedure is illustrated by means of examples for didactic purposes.

11.1 Discrete Systems

The analytical approach illustrated in the preceding chapters gives satisfactory results when simple and clear-cut schemes, necessary for the correct application of the mathematical tools, can adequately interpret the real conditions of the river and the mechanism of pollutant injection. Very often, reality is so complex that the analytical algorithms are not able to take into account all the details. In this case, the analytical approach would require formulations very difficult to deal with, even with the use of the most advanced computing facilities.

On the other hand, it should be taken into consideration that also in the analytical approach, the computers work on a sequence of numbers, to which the mathematical formulations are applied in a "discrete" form. This paradigm of work has favoured the development of mathematical theories based on a sequence of steps with fixed value. The "numerical" or "digital" approach is now able to face many scientific fields, transforming the continuum into a discontinuous domain. Moreover, the methods developed in this context, working with the support of the powerful computers, are able to deal with very complex conceptual simulation mechanisms, and therefore, there are increased possibilities to solve problems that were considered with no solution in the past.

M. Benedini and G. Tsakiris, *Water Quality Modelling for Rivers and Streams*,
Water Science and Technology Library 70, DOI 10.1007/978-94-007-5509-3_11,

This type of processes is applied also to various water problems, bringing in their full power and opening the possibility to tackle and interpret many new aspects of all related problems.

So far, the preceding chapters were developed following the assumption that the water body is a continuum, in which all the terms are identified in space and time and the dynamic field can be also interpreted by a continuous function, as presented by the resulting expressions based on the integration of the fundamental differential equations.

The recent progress in numerical simulation requires that all the expressions proposed so far have to be revised and adapted to the numerical procedures. The water body and the relevant phenomena should be therefore "discretised" and interpreted by means of a set of finite entities, each one having a defined value of the descriptive terms.

Fundamental point of the numerical approach is the assumption that the water body is made up by a sequence of several segments (or "boxes"), according to which the differential equations can be transformed into finite expressions.

Several approaches are available in the scientific literature. In the following paragraphs, the scheme originally proposed by Thomann (1963) is considered, which seems as the most efficient illustration of the approach. The scheme concerns an application to a river, in which only the longitudinal side is of importance, while the dynamic and quality terms are assumed to be constant over the entire cross section.

It is a common and frequent case in which the pollution propagates on very long stretches of the river. To face this case, the river is interpreted as a sequence of reaches, sometimes several kilometres long, having uniform average geometric and hydraulic characteristics. The reaches are limited by the initial and final demarcation faces, which correspond to cross sections easily identifiable in the river (e.g. a bridge, a junction with a tributary or other characteristic cross sections).

Figure 11.1 refers to a study carried out in the early 1970s on the River Tiber and its tributary Aniene crossing the urban area of Rome (Italy), for the development of a water quality model (Chi and Harrington 1973).

The Tiber, from the Castel Giubileo bridge down to the mouth in the Tyrrhenian Sea, is represented by means of 14 reaches, while the Aniene, from Tivoli to its junction with Tiber, is represented by means of 8 reaches, identified with the numbers from 15 to 22 (Fig. 11.2).

Every reach has its own characteristic hydraulic and water quality terms, which are normally assumed as the mean of the values measured or estimated at the initial and final face of the same reach, respectively.

With reference to Fig. 11.3, in an elementary finite portion of the river, the aforesaid processes of advection, dispersion, "sink" and "sources", as well as those of nonconservative pollutants, can be expressed following the balance of the pollutants entering and leaving the reach. As in the figure, $Q_{i-1,i}$ is the flow passing through the demarcation cross section between the elementary reaches $i-1$ and i, and $C_{i-1,i}$ is the average pollutant concentration at the same cross section; W_i and F_i are, respectively, the source and sink relevant to the box.

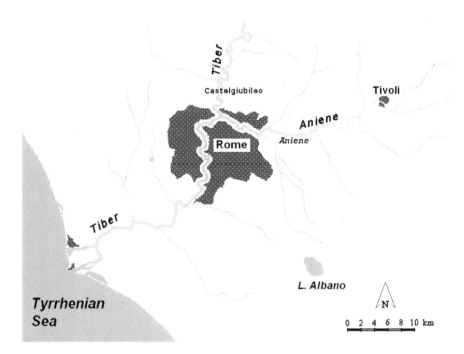

Fig. 11.1 The River Tiber and its main tributary Aniene in the area of Rome (Italy)

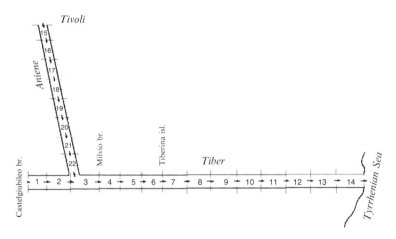

Fig. 11.2 Representation of Tiber and Aniene rivers shown in Fig. 11.1 by means of 22 reaches for the development of a water quality model

The transported pollutant is, therefore, the mass crossing that section due to the combination of the various processes that have been described in the preceding chapters for the continuous case.

The various mechanisms responsible for pollution transport are interpreted as follows.

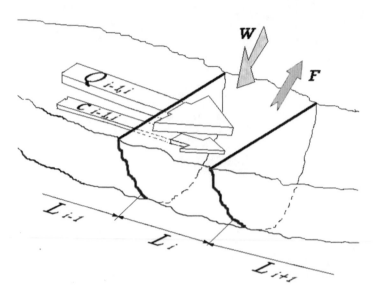

Fig. 11.3 Interpretation of the river by means of discrete segments ("boxes")

11.1.1 Advection

The pollutant mass inflow in the unit time can be expressed as

$$I_i^a = Q_{i-1,i}\, C_{i-1,i}$$

where the superscript a denotes the advection process.

Similarly, at the exit of the same river reach, the mass outflow from reach i and entering the reach $i + 1$ is

$$U_i^a = Q_{i,i+1}\, C_{i,i+1}$$

The net balance due to the advection, in the unit time, is

$$I_i^a - U_i^a\,. \tag{11.1}$$

with dimensions $[MT^{-1}]$.

11.1.2 Dispersion

In line with the application of Fick law, the pollutant mass entering reach i through the demarcation cross section between the reaches $i - 1$ and i can be expressed as

Fig. 11.4 The characteristic
terms for the interpretation of
the dispersion transport of
pollutants between two
successive boxes

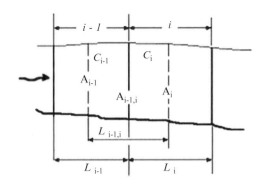

$$I_i^{\mathrm{d}} = \frac{E_{i-1,i}\, A_{i-1,i}\, (C_{i-1} - C_i)}{L_{i-1,i}} \tag{11.2}$$

where superscript d stems from dispersion, $A_{i-1,i}$ is the demarcation cross-sectional area and $L_{i-1,i}$ is a convenient longitudinal distance over which the concentration changes.

If the river characteristics vary within acceptable limits, as presented in Fig. 11.4, it may be assumed that the representative cross sections $A_{i-i,i}$ and A_i are located in the gravity centre of reach $i-1$ and reach i, respectively, or in the middle of the corresponding length of the reach. Therefore, the following equation can be written as

$$L_{i-1,i} = \frac{L_{i-1} + L_i}{2} \tag{11.3}$$

and similarly

$$A_{i-1,i} = \frac{A_{i-1} + A_i}{2} \tag{11.4}$$

Based on the above, the Eq. (11.2) becomes

$$I_i^{\mathrm{d}} = \frac{E_{i-1,i}\, (A_{i-1} + A_i)(C_{i-1} - C_i)}{L_{i-1} + L_i} \tag{11.5}$$

or

$$I_i^{\mathrm{d}} = \bar{E}_{i-1,i}\, (C_{i-1} - C_i) \tag{11.6}$$

where

$$\bar{E}_{i-1,i} = E_{i-1,i}\, \frac{A_{i-1} + A_i}{L_{i-1} + L_i} \tag{11.7}$$

is the "exchange coefficient" between stretch $i - 1$ and i. Its dimensions are $[L^3 \, T^{-1}]$.

Similarly, the mass outflow is

$$U_i^d = \bar{E}_{i,i+1} \, (C_i - C_{i+1}) \tag{11.8}$$

and the net balance due to dispersion is

$$I_i^d - U_i^d \tag{11.9}$$

expressed in terms of $[MT^{-1}]$.

11.1.3   Nonconservative Pollutants

The concepts presented in Chap. 4, valid for an infinitesimal particle, can be extended to the total volume, Y_i, of the stretch, and this contribution can be put as

$$\pm KY_iC_i$$

The double sign means that the reaction can increase or decrease the original pollutant concentration.

11.1.4   Sinks and Sources

The concepts dealt within Chap. 3 are also valid, considering the algebraic sum of all the contributions in the reach. As in Fig. 11.3, the external contribution S is split into two components, the entering one, W, and the outgoing F.

11.1.5   The Demarcation Cross Sections of the Reaches

Very often, the known values of flow and water quality do not refer to the demarcation cross section between two adjacent reaches, but to a cross section inside the single reach, where it is easier to perform a measurement. This entails that it is necessary to find a way for the evaluation of $Q_{i-1,i}$, $Q_{i,i+1}$, $C_{i-1,i}$ and $C_{i,i+1}$.

The following considerations are relevant to the concentration but can also be applied to the flow. There is no restriction to assume that C is continuous and decreases along the water body, as illustrated in Fig. 11.5. The known values are at the cross sections ς_{i-1}, ς_i and ς_{i+1}, concerning the i-th reach. This suggests that the

Fig. 11.5 Concentration values for the river stretches

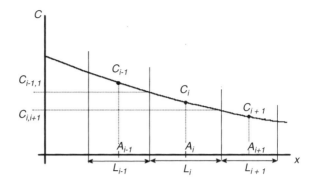

concentration can be calculated by interpolation adopting a "weighting factor" P $(0 \leq P \leq 1)$.

For the upstream cross section

$$C_{i-1,i} = C_i + P_{i-1,i}\left(C_{i-1} - C_i\right) \tag{11.10}$$

and, similarly, for the downstream one

$$C_{i,i+1} = C_{i+1} + P_{i,i+1}\left(C_i - C_{i+1}\right) \tag{11.11}$$

11.1.6 The Fundamental Equation

Combining the previous numerical formulations, in the same way as done for the infinitesimal particle in Chap. 4, after some algebraic manipulations, the general finite-term equation for the nonconservative pollutant can be written as

$$Y_i \frac{\partial C_i}{\partial t} = Q_{i-1,i}\left[P_{i-1,i}\, C_{i-1} + \left(1 - P_{i-1,i}\right) C_i\right] + \bar{E}_{i-1,i}\left(C_{i-1} - C_i\right)$$
$$- Q_{i,i+1}\left[P_{i,i+1}\, C_i + \left(1 - P_{i,i+1}\right) C_{i+1}\right] - \bar{E}_{i,i+1}\left(C_i - C_{i+1}\right) \pm kY_iC_i \pm w_i$$
$$\tag{11.12}$$

The value of P can be defined from the experimental values of C along the river stretch. Assuming $P = 1$, it follows

$$C_{i-1,i} = C_{i-1}$$

that is to say that the concentration has a "jump" passing from a reach to the next one and remains constant for the whole length of the reach. This is another example of the *plug flow* approach, already mentioned in Chap. 10. Similar considerations

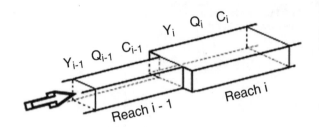

Fig. 11.6 The main characteristics of the "plug flow" approach

hold also for the other terms of the problem, in a way that the "boxes" representing the reality are in fact independent from each other, as illustrated in Fig. 11.6.

This assumption based on the predominance of advection transport leads to further simplified computations.

The assumption of steady-state conditions is also frequently assumed. Then, Eq. (11.12) becomes

$$Q_{i-1,i}\left[P_{i-1,i}\,C_{i-1} + \left(1 - P_{i-1,i}\right)C_i\right] + \bar{E}_{i-1,i}\left(C_{i-1} - C_i\right) - Q_{i,i+1}$$
$$\left[P_{i,i+1}\,C_i + \left(1 - P_{i,i+1}\right)C_{i+1}\right] - \bar{E}_{i,i+1}\left(C_i - C_{i+1}\right) - kY_iC_i \pm w_i = 0$$

(11.13)

in which the pollutant concentration is the current variable.

11.1.7   Combined Processes

Equation (11.12) or (11.13) allows for a combined process to be considered, like in the common case of BOD and DO interaction. After putting into evidence the concentration, the BOD equation can be written as

$$\left[Q_{i-1,i}P_{i-1,i} + \bar{E}_{i-1,i}\right]C_{i-1} + \left[Q_{i-1,i}\left(1 - P_{i-1,i}\right) - Q_{i,i+1}P_{i,i+1} - \bar{E}_{i-1,i} - \bar{E}_{i,i+1} - KY_i\right]C_i +$$
$$\left[-Q_{i,i+1}\left(1 - P_{i,i+1}\right) + \bar{E}_{i,i+1}\right]C_{i+1} + f_i = 0$$

(11.14a)

where C_i is the BOD concentration in reach i and K is the deoxygenation coefficient.

For DO, according to Chap. 6, the equation is

$$\left[Q_{i-1,i}P_{i-1,i} + \bar{E}_{i-1,i}\right]s_{i-1} + \left[Q_{i-1,i}\left(1 - P_{i-1,i}\right) - Q_{i,i+1}P_{i,i+1} - \bar{E}_{i-1,i} - \bar{E}_{i,i+1} - K_aY_i\right]s_i +$$
$$\left[-Q_{i,i+1}\left(1 - P_{i,i+1}\right) + \bar{E}_{i,i+1}\right]s_{i+1} + K_aY_iS_s - KY_iC_i \pm g_i = 0$$

(11.14b)

where s_i is the DO concentration in reach i, K_a the reaeration coefficient and taking into account the saturation concentration, S_s, of DO, supposed to be constant for all the river stretches that are analysed.

11.2   An Example

The simple example considered in the following paragraphs can assist the reader to understand the concepts introduced earlier in this chapter.

A concentrated amount of BOD, released by a polluted area, is injected into a river, in which a natural waterfall is located upstream of the point of pollutant injection, as shown in the plan of Fig. 11.7.

For the problem concerning the pollution propagation, the stretch can be considered as a sequence of four reaches having constant rectangular cross section and slope. According to the empirical criterion of BOD evaluation (Chap. 6), the pollutant load of 220.00 g/s is due to the discharge of a resident population of 200,000 inhabitants, combined with the equivalent population of food industries (150 employees), alcohol beverage factories (50 employees), oil mills (50 employees), chemical and synthetic fabrics (250 employees each) and metallurgical factories (100 employees), based on the assumption of the per capita BOD discharge of 70 g/day.

The longitudinal profile in Fig. 11.8 shows the hydraulic characteristics, while all the data of the problem are summarised in Table 11.1.

The river behaves as plug flow, and, as a further simplification (which does not affect the inner mechanisms of the procedure), it is assumed that there is no dispersion, and the pollutant transport is performed only by advection. Following such assumptions, after putting $P = 1$ and $E = 0$, Eqs. (11.14a) and (11.14b) can be further simplified leading to

$$Q_{i-1} C_{i-1} - Q_i C_i - KY_i c_i + f_i = 0 \qquad (11.15)$$

$$s_{i-1} Q_{i-1} - s_i(Q_i + K_a Y_i) + K_a Y_i S_s - KY_i C_i + g_i = 0 \qquad (11.16)$$

Concerning Eq. (11.16), it is a common practice to express it in terms of DO deficit, following the considerations analysed in Chap. 7.

Fig. 11.7 The river stretch considered in the example

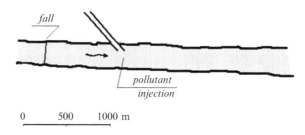

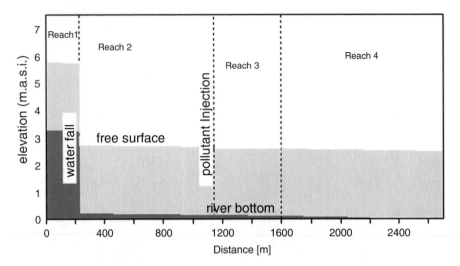

Fig. 11.8 Longitudinal profile of the river stretch

Table 11.1 Data of the problem

	Unit	Reach 1	Reach 2	Reach 3	Reach 4
Length	m	229	914	457	1,143
Volume	m³	8,000	32,000	18,885	47,211
Slope	–	0.00009	0.00009	0.00009	0.00009
Flow	m³/s	30	30	34	34
Depth	m	2.5	2.5	2.43	2.43
Width	m	14	14	17	17
Velocity	m/s	0.86	0.86	0.82	0.82
Velocity head	m	0.04	0.04	0.03	0.03
Depth of the fall		3 m			
Roughness coefficient		$0.017 \text{ m}^{-1/3} \text{ s}$			
Pollutant flow		4.00 m³/s			
Pollutant load (BOD)		220.00 g/s			
Water temperature		20 °C			
Initial BOD concentration		5.00 g/m³			
Initial DO concentration		7.22 g/m³			
Deoxygenation coefficient		$0.00000463 \text{ s}^{-1}$			
Reaeration coefficient		$0.00001042 \text{ s}^{-1}$			

With the water temperature of 20 °C (summer conditions), according to expression (6.11) of Chap. 6, the concentration of dissolved oxygen saturation is $S_s = 9.022 \text{ g/m}^3$, while expression (6.9), for the fall depth of 3.00 m at the beginning of reach 2, gives a decrease of oxygen deficit of $\sigma_2 = 0.190\, \sigma_1$.

Then the deficit is

$$\sigma_I = S_s - s_i \qquad (11.17)$$

and Eq. (11.6) becomes

$$Q_{i-1}\,\sigma_{i-1} - (Q_i + K_{DO}Y_i)\sigma_i + S(Q_{i+1} - Q_i) - K_{BOD}Y_iC_i - g_i = 0 \qquad (11.18)$$

Expressing the DO in terms of deficit simplifies not only the calculation but also the introduction of sinks and of sources, which are normally expressed as deficit decrease.

With the specific data of the problem, Eq. (11.15) gives rise to the following system of linear equations:

$$
\begin{array}{rrrrr}
-30.04 \ \ C_1 & & & & = -150.00 \\
30.00 \ \ C_1 & -30.15 \ \ C_2 & & & = \ \ \ \ 0.00 \\
& 34.00 \ \ C_2 & -34.09 \ \ C_3 & & = -220.00 \\
& & 34.00 \ \ C_3 & -34.22 \ \ C_4 & = \ \ \ \ 0.00
\end{array}
$$

Following the general theory of linear algebra, the system can be written in vector form as

$$\mathbf{BC} = \mathbf{F} \qquad (11.19)$$

in which $\mathbf{B}$ is *the matrix of the coefficients of unknowns,* $\mathbf{C}$ the *vector of variables (unknowns)* and $\mathbf{F}$ the *vector of known terms,* namely,

$$
\mathbf{B} = \begin{vmatrix}
-30.04 & 0.00 & 0.00 & 0.00 \\
30.00 & -30.15 & 0.00 & 0.00 \\
0.00 & 34.00 & -34.09 & 0.00 \\
0.00 & 0.00 & 34.00 & -34.22
\end{vmatrix}
$$

$$\mathbf{C} = \mathbf{C}(\,C_1,\ C_2,\ C_3,\ C_4)$$
$$\mathbf{F} = \mathbf{F}(-150.00,\ 0.00, -220.00,\ 0.00)$$

The solution of (11.18) is achieved as

$$\mathbf{C} = \mathbf{B}^{-1}\mathbf{F}$$

in which $\mathbf{B}^{-1}$ is the inverse matrix of $\mathbf{B}$

$$
\mathbf{B}^{-1} = \begin{vmatrix}
-0.033 & 0.00 & 0.00 & 0.00 \\
-0.033 & -0.033 & 0.00 & 0.00 \\
-0.029 & -0.029 & -0.029 & 0.00 \\
-0.029 & -0.029 & -0.029 & -0.029
\end{vmatrix}
$$

Therefore,

$$C_1 = \ \ 4.99 \ \text{g/m}^3$$
$$C_2 = \ \ 4.97 \ \text{g/m}^3$$
$$C_3 = 10.83 \ \text{g/m}^3$$
$$C_4 = 10.76 \ \text{g/m}^3$$

Slightly more complex is the calculation of DO by means of Eq. (11.18). It should be remembered that the equation relevant to reach 2 contains the source of DO due to the waterfall; then, with the data of Table 11.1 and remembering (11.17), such an equation is

$$30.0\sigma_1(1 - 0.190) - 30.41\sigma_2 + 0.736 = 0$$

Applied to the four reaches of the river stretch, Eq. (11.18) gives rise to the following linear system:

$$
\begin{array}{rlrlrlrlcr}
-30.10 & \sigma_1 &   &   &   &   &   &   & = & -54.32 \\
 24.28 & \sigma_1 & -30.41 & \sigma_2 &   &   &   &   & = & -0.74 \\
        &   & 30.00 & \sigma_2 & -34.24 & \sigma_3 &   &   & = & -37.03 \\
        &   &   &   & 34.00 & \sigma_3 & -34.60 & \sigma_4 & = & -2.65
\end{array}
$$

with the matrix of the coefficients of unknowns

$$
D = \begin{vmatrix}
-30.10 & 0.00 & 0.00 & 0.00 \\
 24.28 & -30.41 & 0.00 & 0.00 \\
  0.00 & 30.00 & -34.24 & 0.00 \\
  0.00 & 0.00 & 34.00 & -34.60
\end{vmatrix}
$$

The system can be written in vector form

$$DS = G \tag{11.20}$$

in which the *vector of variables* S and the *vector of known terms* G, for the specific case, are

$$S = S(\sigma_1, \ \sigma_2, \ \sigma_3, \ \sigma_4)$$
$$G = G(-54.32, -0.74, -37.04, -2.35)$$

The latter expression contains the contribution of the deoxygenation values of BOD previously calculated. The solution is

$$S = D^{-1}G$$

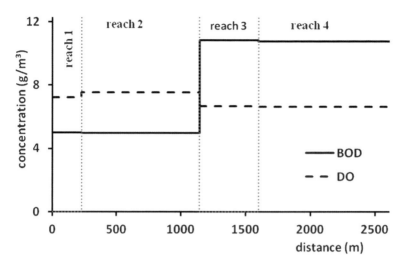

Fig. 11.9 BOD and DO values in the river stretch

where

$$\mathbf{D}^{-1} = \begin{vmatrix} -0.03 & 0.00 & 0.00 & 0.00 \\ -0.03 & -0.03 & 0.00 & 0.00 \\ -0.02 & -0.03 & -0.03 & 0.00 \\ -0.02 & -0.03 & -0.03 & -0.03 \end{vmatrix}$$

that is $\mathbf{D}^{-1}$ inverse matrix of $\mathbf{D}$. The final results are

$$\sigma_1 = 1.80\,\mathrm{g/m^3}$$
$$\sigma_2 = 1.47\,\mathrm{g/m^3}$$
$$\sigma_3 = 2.37\,\mathrm{g/m^3}$$
$$\sigma_4 = 2.39\,\mathrm{g/m^3}$$

or, transformed into the real value of DO,

$$S_1 = 7.22\,\mathrm{g/m^3}$$
$$S_2 = 7.56\,\mathrm{g/m^3}$$
$$S_3 = 6.66\,\mathrm{g/m^3}$$
$$S_4 = 6.63\,\mathrm{g/m^3}$$

To conclude the example, the results of the application are shown in Fig. 11.9.

The figure shows that the model, in spite of the many approximations, is able to give a clear picture of the pollutant behaviour in the river, comparable to the behaviour illustrated for the analytical approach in the preceding chapters.

The procedure illustrated with the example above stresses the role of the matrices, which allow, for a given input, a particular output to be obtained. This way of proceeding belongs to the "input-output modelling", developed in the early 1970s as a tool of system analysis. Many interesting applications have been presented in the past, the most famous of which is that related to the river Delaware in the United States.

The approach illustrated can be completed inserting more terms (like the dispersion) and assuming a more detailed representation of the river geometry. In line with the results of the applications shown in Chap. 10, it is worthy to notice that the presence of BOD in water affects the DO, but not vice versa. In other words, the decay of BOD does not depend on the initial content of oxygen, but, as expressed by the relevant equation, it complies with the chemical and biological rules governing the transformation of bacteria and organic matter.

The response of the model, as plotted in the Fig. 11.9, is in form of "steps" of constant values, covering the entire length of the river reach in which the concentration is considered. This type of solution is typical of the discontinuous approach adopted for the model in finite terms. Replacing the continuous approach of the analytical model entails also a new way of interpreting the reality. Consequently, assuming a constant concentration value for a long river reach may be inappropriate, because it is expected to vary continuously, and, under this viewpoint, the discrete approach can be considered as less precise than the analytical one already described in the preceding Chapters.

As a general consideration, it could be reasonable to assume that the calculated constant value is referred to the "gravity centre" of the reach, which can be approximated in a cross section located at the mid-length and represented by an isolated point. The assumption allows the concentration values to be joined by means of a continuous line, giving a more realistic picture to the reality, especially if the length of the river reach is small.

11.3   Additional Comments

The model developed by Thomann in the 1960s is one of the first examples of numerical application in water quality problems and opened the way to further steps in numerical modelling, challenging the mathematicians and programmers to devise more comprehensive and powerful tools that can take advantage from the computer progress.

Several applications of the Thomann model have been performed leading to satisfactory results. An interesting case was that of the urban stretches of the river Tiber in the area of Rome, already mentioned. The results were in line with the corresponding experimental data derived from direct measurements. Figure 11.10

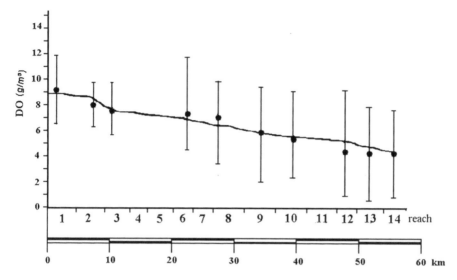

Fig. 11.10 Application of the Thomann model to the urban stretch of the river Tiber (see Figs. 11.1 and 11.2). The model output is indicated as a *solid line*, while the measurements are indicated with the mean value and the range of values

shows the DO values in the main stretch of the river, in which the model output has been compared with the measurements carried out in the stream.

It is particularly worthy to notice that the variability of the experimental values can be very large, while the mean values remain very close to the model output. This can be considered as an encouraging proof of the intrinsic power of the numerical models. Nevertheless, the Thomann model suffers from several drawbacks, due principally to the state of the art in numerical and computing facilities existing in that period.

The first drawback is in the assumption of the steady-state conditions of the river, which allow the model to take only a snapshot of the reality, excluding any possibility to simulate how the pollutant behaves after being discharged in the stream. This assumption may be inconsistent with the natural behaviour of the river and does not take into account that the pollutant concentration can vary within the length of each reach, as soon as the pollutant comes in contact with water.

A second negative aspect is in the "rigid" scheme adopted to interpret the reality, which does not permit any consideration except the one-dimensional approach.

In principle, all these disadvantages may be overcome, by adopting, for instance, shorter reaches in long river stretches and considering the time dependence as a sequence of several steady-state situations. This trick would only in part abate the weakness of the approach, because it introduces useless complexity in the application, without increasing the model reliability.

The innovation in numerical analysis and the progress in computing can now provide tools that allow a better interpretation of the real conditions, achieving more reliable results and saving computing time. Since the Thomann model, several approaches have been proposed, and many new ones are progressively introduced by the scientific community.

It seems wise to insert these considerations in the more general context of water management problems recalling the history of water quality models. In fact, water quality modelling has evolved since the late twentieth century, following the way the water quality problems were evaluated by the society and in accordance with the progress of the computational capabilities, both in terms of hardware and software. Originally, the models consisted of simple formulations inferred from experimental works in the field, able to correlate the observed values with some fundamental terms of the stream. Due to the limitation of the computational tools, the main applications were restricted to linear kinetics, simple geometries and steady-state conditions. These formulations contributed to develop the fundamental differential equation that was successfully transformed into the analytical models described in the preceding chapters.

As soon as the digital computers became widely available, the formulations could be interpreted by means of the numerical approach, which allowed more complicated analyses to be carried out, including complex system geometries, kinetics and time-variable simulation processes. Tools originally developed in the field of "operational research" and "system analysis" were applied to simulate the effect of treatment alternatives, including cost evaluations, in the framework of an environmental remediation planning.

Adoption of numerical procedures can benefit from a large number of mathematical tools, some of which are still underdevelopment, associated with the tremendous progress of the computing facilities.

The following chapters describe some of the most common procedures, with a sufficient number of references for the interested reader.

References

Chi T, Harrington JJ (1973) A water quality management model for the Tiber Basin. Italian Water Research Institute, Technical report of META SYSTEM INC., Cambridge, MA

Thomann RV (1963) Mathematical model for dissolved oxygen. J Sanit Eng Div ASCE SA5:1–30

Chapter 12
Progress in Numerical Modelling: The Finite Difference Method

Abstract Some partial differential equations (PDE) that describe several hydrodynamic phenomena do not have analytical solution, and hence they can only be solved by numerical methods. The finite difference method (FDM) is traditionally the most efficient way for numerical integration of PDEs, now empowered by the availability of computing facilities. Applied to the water quality models, the FDM has been developed according to several numerical schemes which are useful in practical applications. Some of these numerical schemes are described in this chapter, focussing on their particular characteristics, in order to provide guidance for the most frequently encountered cases of pollution in rivers and streams.

12.1 Outlines of the Most Common Numerical Methods

After being discretised, the general differential equation of pollution transport is replaced by an expression of finite terms, both in space, with intervals of prefixed size, Δx, Δy and Δz, along the coordinate directions, and in time, with a Δt interval.

The problem is now to move to the larger size of the entire water body under examination. Such an action corresponds to the integration process suited for the differential equation.

Manipulation of finite terms is an old challenge in mathematics, which involves repetitive tasks, as the problem is normally brought to the solution of linear equations, or linear equation systems, with many steps and variables. The availability of computers has permitted great improvements of these procedures, and many mathematical tools have been developed accordingly (Eheart 2006; Fernandez and Karney 2001).

At the present time, generally speaking, the numerical methods can be grouped into three broad categories, namely, the finite difference method (FDM), the finite element method (FEM) and the finite volume method (FVM). Some of the substantial characteristics of these methods are outlined in the following pages.

M. Benedini and G. Tsakiris, *Water Quality Modelling for Rivers and Streams*,
Water Science and Technology Library 70, DOI 10.1007/978-94-007-5509-3_12,

12.2   The Finite Difference Method (FDM)

It is the simplest and original method for solving a differential equation and consists of replacing the infinitesimal terms with finite steps of convenient size. All the steps should be equal, even though some recent studies have proposed steps of different size that can better interpret the reality under examination.

The adoption of discontinuous finite terms does not necessarily maintain all the characteristics of the original differential expression, because these terms cannot take into account all the details that only a continuous approach can adequately describe. Moreover, even under the mathematical perspective, there are aspects which are difficult to be handled. As a matter of fact, the solution is always approximated.

A *true solution* is in practice a meaningless entity, as all the approaches, also the analytical one, suffer from some kind of approximation. Nevertheless, a discrete interpretation is currently characterised by a *truncation error*, which can be appreciated comparing several solutions obtained by means of different procedures, and reflects, implicitly, the fact that a discontinuous step cannot be viewed as a finite part of the infinite.

Assuming, for example, that f is a function of the independent variable x, its partial derivative at point x_0 is

$$\frac{\partial f}{\partial x} = \lim_{\Delta x \to 0} \frac{f(x_0 + \Delta x) - f(x_0)}{\Delta x} \tag{12.1}$$

If the function f is defined at point x_0, then at point $x_0 + \Delta x$, it can be approximated by Taylor series as follows:

$$f(x_0 + \Delta x) = f(x_0) + \sum_1^n \frac{\Delta x^n \partial^n f}{n! \partial x^n} = f(x_0) + \Delta x \frac{\partial f}{\partial x} + \Delta x^2 \frac{\partial^2 f}{2 \partial x^2} + \dots \tag{12.2}$$

Keeping only the first term of the right-hand side, the latter equation gives

$$\frac{\partial f}{\partial x} = \frac{f(x_0 + \Delta x) - f(x_0)}{\Delta x} \tag{12.3}$$

From the above equation, for a finite interval Δx with small value, the partial derivative of function f can be determined with an approximation by a finite difference. The other terms that are neglected from the Taylor series constitute the truncation error described above. The other category of errors encountered in numerical methods makes up the *rounding errors*.

Essential for the application of the FDM is the way by which the water body is discretised in finite intervals with respect to space and time. There are various forms of doing this, as illustrated in the following pages. They all have their own peculiarities and can lead under certain circumstances to solutions that cannot be accepted, because they depart far from a real interpretation of the examined phenomenon. There are mathematical investigations aiming at criteria and recommendations for an approach that can avoid unacceptable errors (Stamou 1991). In a quite general perspective, any application of the FDM requires an accurate analysis of the real problem to be examined.

At the present time, the software market offers various packages, suitable for application in a large variety of practical cases, which seem capable for keeping the errors within an acceptable range. Obviously, these packages take into account that all the mathematical representations of the pollution transport in a river are affected by the difficulty of having precise evaluations of the various physical entities involved.

12.3 Basic Concepts of the Numerical Approach

To understand the basic concepts of the numerical approach, in the one-dimensional case, for an advective-dispersive-reactive fluid, it is convenient to consider a representation in the x, t plane, as presented in Fig. 12.1.

The six points in the grid represent the situation of three cross sections, respectively, at the $x - 1$, x and $x + 1$ coordinate and at the two time instants t and $t + 1$. Each point is characterised by a specific value of velocity, U, in the x-direction, and by a pollutant concentration C.

Interpretation of pollution transport in discrete terms implies rewriting Eq. (4.8) of Chap. 4 in a finite form.

If Δt is the interval between the two time instants and Δx is the distance between the two cross sections at the $x - 1$ and x location and between x and $x + 1$, the partial derivative of concentration with respect to time can be written as

$$\frac{\partial C}{\partial t} \rightarrow \frac{1}{\Delta t} \left[\frac{C_x^{t+1} + C_{x+1}^{t+1}}{2} - \frac{C_x^t + C_{x+1}^t}{2} \right] \tag{12.4}$$

assuming at every instant the concentration as the average values of two contiguous points in the grid.

For the advection process, the entering mass through the cross section at $x - 1$ can be put as

$$\frac{C_x^t U_x^t + C_x^{t+1} U_x^{t+1}}{2}$$

or the average value in the considered time interval.

Similarly, the mass leaving through the cross section at $x + 1$

Fig. 12.1 The x,t grid for the numerical interpretation

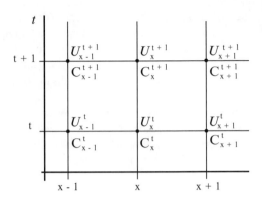

$$\frac{C_{x+1}^{t} U_{x+1}^{t} + C_{x+1}^{t+1} U_{x+1}^{t+1}}{2}$$

The advection term, with the proper sign, is therefore

$$-\frac{\partial CU}{\partial x} \rightarrow \frac{1}{\Delta x}\left[\frac{C_x^t U_x^t + C_x^{t+1} U_x^{t+1}}{2} - \frac{C_{x+1}^t U_{x+1}^t + C_{x+1}^{t+1} U_{x+1}^{t+1}}{2}\right] \qquad (12.5)$$

For the dispersion process, the entering mass through the cross section at x is

$$\frac{E}{\Delta x}\left[\frac{C_x^t + C_x^{t+1}}{2} - \frac{C_{x-1}^t + C_{x-1}^{t+1}}{2}\right]$$

and that leaving through the section at $x + 1$

$$\frac{E}{\Delta x}\left[\frac{C_{x+1}^t + C_{x+1}^{t+1}}{2} - \frac{C_x^t + C_x^{t+1}}{2}\right]$$

The dispersion term, after some rearrangement, is therefore

$$\frac{\partial^2 c}{\partial x^2} \rightarrow \frac{E}{2\Delta x^2}\left[C_{x+1}^t + C_{x+1}^{t+1} - 2\left(C_x^t + C_x^{t+1}\right) + C_{x-1}^t + C_{x-1}^{t+1}\right] \qquad (12.6)$$

The reactive term can be expressed as

$$k\frac{C_{x+1}^t + C_{x+1}^{t+1} + C_x^t + C_x^{t+1}}{4} \qquad (12.7)$$

Combining Eqs. (12.4), (12.5), (12.6) and (12.7) and after some convenient rearrangement, the resulting equation becomes

$$\omega_1 C_x^t + \omega_2 C_{x+1}^t + \omega_3 C_x^{t+1} + \omega_4 C_{x+1}^{t+1} + \bar{\omega} = 0 \qquad (12.8)$$

where

$$\omega_1 = \left[\frac{1}{2\Delta t} - \frac{U_x^t}{2\Delta x} - \frac{E}{\Delta x^2} - \frac{k}{4} \right] \qquad (12.9a)$$

$$\omega_2 = \left[\frac{1}{2\Delta t} - \frac{U_{x+1}^t}{2\Delta x} + \frac{E}{2\Delta x^2} - \frac{k}{4} \right] \qquad (12.9b)$$

$$\omega_3 = \left[-\frac{1}{2\Delta t} - \frac{U_x^{t+1}}{2\Delta x} - \frac{E}{\Delta x^2} - \frac{k}{4} \right] \qquad (12.9c)$$

$$\omega_4 = \left[-\frac{1}{2\Delta t} - \frac{U_{x+1}^{t+1}}{2\Delta x} + \frac{E}{2\Delta x^2} - \frac{k}{4} \right] \qquad (12.9d)$$

and

$$\bar{\omega} = \left[\frac{E}{2\Delta x^2} \left(C_{x-1}^t + C_{x-1}^{t+1} \right) \right] \qquad (12.9e)$$

with $\bar{\omega}$ the known term.

The FDM can be applied for a more complex approach, including the effect of dispersion and local injections or subtractions. The structure of Eqs. (12.8) and (12.9a, b, c, d and e) allows to be considered for any term.

Equation (12.8) has to be applied for the various Δx steps into which the river is split and for any Δt time interval considered. This gives rise to a system of algebraic linear equations to be solved by means of the ordinary procedures that can benefit from the mathematical tools available in numerical algebra (Zhihua and Elsworth 1989). The solution gives the concentration values in the river as a function of space and time.

Needless to say that in order to solve Eq. (12.8), suitable boundary and initial conditions are required.

12.4 Numerical Schemes: Stability Criteria

The above mathematical procedure is called *numerical scheme*. The most common finite differences are the forward difference, the backward difference and the centred difference. A combination of these basic finite differences generates a lot of numerical schemes. According to the number of terms which are kept from the

Taylor series, the numerical schemes can be categorised as exhibiting the *first-order accuracy* ($O(\Delta x)$), the *second-order accuracy* ($O(\Delta x^2)$), etc.

Generally, the numerical schemes are distinguished in *implicit* and *explicit* schemes; in the implicit numerical scheme, the variable calculation is performed at the end of the iteration by solving the system of equations by means of several methods (Thomas algorithm, Gauss-Seidel method, etc.). In the explicit numerical schemes, the variable calculation is performed at every step of the computational procedure.

The criteria which should be followed in numerical schemes are (a) consistency, (b) stability, (c) convergence, (d) conservation and (e) boundedness.

The discretisation of PDEs can encounter some mathematical adversity, especially in the six-point approach outlined in Eq. (12.8). The main undesirable aspect is a solution $C(x, y, z; t)$ unstable in space and time, assuming values not justifiable in the physical reality. There are terms that do not follow a reasonable sequence, with discontinuous values, sometimes negative, which cannot be interpreted in respect of the real phenomenon under consideration.

Generally, the implicit numerical schemes are stable in any case, while the explicit numerical schemes are stable only under certain conditions. In the following paragraphs, the instability of numerical schemes will be analysed.

One of the first causes of instability in explicit numerical schemes lies in the size of the discretised space, which must comply with the mechanism of advection that considers the pollution transport from one step to the next one. In the one-dimension case, if the length of the space step Δx is smaller than the term $U\Delta t$, the pollutant transport occurs outside the box taken into consideration, unlike the assumptions of Sect. 12.3. The quantity

$$C_{ou} = \frac{U\Delta t}{\Delta x} \tag{12.10}$$

is known as *Courant number*; generally, for stability, it should be

$$C_{ou} \leq 1 \tag{12.11}$$

The above condition is known as *Courant-Friedrichs-Lewy condition* (CFL condition).

The second reason of instability should be sought in the structure itself of the fundamental differential equation of pollution transport. In its mathematical form, it is made up by a *hyperbolic* part, resulting from the advection terms, and by a *parabolic* part, resulting from the diffusion terms.

In numerical analysis, these terms require a different approach, which cannot be always merged into a unique procedure. The integration process is heavily conditioned by the simultaneous presence of the two types of terms.

These concepts are summarised in the *Péclet number*

$$Pe = \frac{U\Delta x}{E} \tag{12.12}$$

which represents the ratio between the advection and dispersion transport.

Numerous research efforts have highlighted the relations between the Péclet number and the stability of the expected solution. A "rule of thumb" has pointed out, as a necessary but not sufficient condition, that the Péclet number must be less than 2.

More accurate analyses have focussed on the interaction between the geometrical terms and the physical aspects of the problem. Another dimensionless term has been identified, namely, the *diffusion number*

$$\Lambda = \frac{E \Delta t}{\Delta x^2} \qquad (12.13)$$

pointing out that the solution stability means that the errors should not be amplified as far as the computation proceeds. To avoid such an event, a simple empirical rule suggests that Λ should be less than 0.2.

It is recommended to comply with the above criteria, even though in several numerical schemes, they are insufficient to secure a stable solution.

A comprehensive analysis of the stability problem is presented by Chapra (1997).

12.5   Boundary Conditions

Suitable boundary and initial conditions are prerequisite in order to solve the PDEs by means of numerical methods. According to the above paragraphs, initial conditions reflect the values at $x = 0$ and $t = 0$ of the variables in the computational field.

The initial value $C(0,0)$, denoted for simplicity C_0, is known, at least in the very common cases of pollutant injection at the beginning of the stream taken into consideration. Similarly, the initial value $U(0,0)$, denoted U_0, is known, since the dynamic characteristics of the whole stream are always assumed to be preliminarily determined.

Different considerations hold for the final reaches of the stream, theoretically indicated as $C(\infty,\infty)$, but in practice identified at a reasonable distance and at a convenient time as to consider that a stable situation has been reached.

Concerning the behaviour of a pollutant in a water body, there are in practice two different alternatives leading to different interpretations.

The first alternative concerns the case where the stable situation is known a priori. This is the typical case of a continuous injection of a persistent pollutant, for which the reaction is not active and the relevant coefficient k is constantly zero. At a certain distance from the injection and at certain time, the saturation value, C_0, is fully attained.

Such a situation, in which the computed variables are directly determined, interprets the *Dirichlet rule* and can be expressed as

$$C(\infty, \infty) = C_0 \qquad\qquad (12.14)$$

or, better

$$C(m, n) = C_0 \qquad\qquad (12.15)$$

with both m and n sufficiently high as to allow the saturation to be reached.

The other alternative reflects the case where a stable concentration is finally achieved, not necessarily of saturation, which cannot be determined a priori, because it is the combined effect of all the mechanisms that intervene in the pollutant transport, particularly the reaction. This case, where the variables are determined by a known derivative, is known as *Neumann rule*. The final downstream condition is in the form

$$\frac{\partial C}{\partial t} = 0 \quad \frac{\partial C}{\partial x} = 0 \qquad\qquad (12.16)$$

and practically

$$C(m, n) = C(m - 1, n - 1) \qquad\qquad (12.17)$$

with m and n sufficiently high.

As a conclusion, the adoption of the boundary conditions follows an accurate analysis of both the stream and pollutant characteristics.

12.6 An Example: Pure Advection Transport

In mathematics the finite difference is at large the classical method of solving numerically the PDEs. Equation (12.8) and its associated equations are clearly one possible way to introduce the method. There is now the necessity to "vest" the equation with the appropriate data and to find a way to achieve a solution, or, in other words, to obtain an expression able to interpret the behaviour of the entire water body under consideration, in its global aspects.

The essential steps of the way to achieve a solution are illustrated in the following example.

For illustration purposes, a very simple case is considered. In a rectilinear channel, the pollution transport is assumed to be only due to advection. The problem is assumed one-dimensional, and a pollutant is injected at the cross section $x = 0$, where its concentration is known. There are no other local injections.

Table 12.1 Data of the problem

Term	Unit	Value
Δx	m	100
U	m/s	0.6
Δt	s	120
E	m²/s	0
k	s^{-1}	0

The case under examination is one of those already described in Chap. 9. The characteristic data are listed in Table 12.1. It is convenient to assume separate space steps Δx, in each one of them to focus on particular time steps Δt. In the example, five space steps are considered.

Equations (12.9a, b, c, d and e) result in the following values:

$$\omega_1 = 0.0011667$$
$$\omega_2 = 0.0071667$$
$$\omega_3 = -0.0071667$$
$$\omega_4 = -0.0011667$$
$$\bar{\omega} = 0.000$$

For the first spatial step Δx, or for $x = 100.00$ m, Eq. (12.8) becomes the recursive equation for $C(x,t)$

$$11.67\ C(0,t+1) + 71.67\ C(1,t+1) - 71.67\ C(0,t) - 11.67\ C(1,t) = 0 \quad (12.18)$$

in which, in order to make it handier, both the sides have been multiplied by 10,000.

Also in this case, the pollutant concentration is expressed in dimensionless form as a percentage of the original value, $C(0,0)$, at $x = 0.00$ m and $t = 0.00$ s.

At the cross section $x = 0$, the concentration $C(0,t)$ remains indefinitely constant and equal to 100.

The stream is undisturbed for its entire length at the beginning

$$C(x,t) = 0 \quad \text{for} \quad x > 0$$

and before the presence of pollutant can be assessed

$$C(x,0) = 0 \quad \text{for} \quad x \geq 0$$

For the time step $t = 1$, Eq. (12.8) gives

$$71.67\ C(1,1) = 6000.00$$

Table 12.2 Solution for the
first spatial step

$C(1,1)$	83.72
$C(1,2)$	97.35
$C(1,3)$	99.57
$C(1,4)$	99.93
$C(1,5)$	99.99

Table 12.3 Solution for the
second spatial step

$C(2,1)$	0.00
$C(2,2)$	67.87
$C(2,3)$	92.19
$C(2,4)$	98.67
$C(2,5)$	99.71

and likewise for the other four time steps. Then, five linear algebraic equations can be written as follows:

$$
\begin{aligned}
71.67 \quad C(1,1) & & & & & = 6000.00 \\
-11.67 \quad C(1,1) \quad +71.67 \quad C(1,2) & & & & & = 6000.00 \\
-11.67 \quad C(1,2) \quad +71.67 \quad C(1,3) & & & & = 6000.00 \\
-11.67 \quad C(1,3) \quad +71.67 \quad C(1,4) & & & = 6000.00 \\
-11.67 \quad C(1,4) \quad +71.67 \quad C(1,5) & = 6000.00
\end{aligned}
$$

The solution of this linear system is shown in Table 12.2

For the second spatial step, at $x = 200.00$ m, the recursive equation becomes

$$11.67\, C(1,t+1) + 71.67\, C(2,t+1) - 71.67\, C(1,t) - 11.67\, C(2,t) - 11.67\, C(1,t) = 0$$

which, applied to the specific case, gives rise to the following linear system:

$$
\begin{aligned}
71.67 \quad C(2,2) & & & & & = 4864.25 \\
-11.67 \quad C(2,2) \quad +71.67 \quad C(2,3) & & & & & = 5815.11 \\
-11.67 \quad C(2,3) \quad +71.67 \quad C(2,4) & & & & = 5995.79 \\
-11.67 \quad C(2,4) \quad +71.67 \quad C(2,5) & & & = 5995.10
\end{aligned}
$$

the solution of which is presented in Table 12.3.

The illustrated procedure can be extended to the other successive time and space steps. At the end of calculation, after replacing the assumed t and x with the values indicated in Table 12.1, the final result is shown in Fig. 12.2, in which, as usual, the pollutant concentration is plotted with its variability in time and space.

The example illustrated concerns a very simple approach, neglecting some other aspects that can be relevant to the problem of evaluating the river behaviour. Nevertheless, this approach is very often adopted for a first-glance evaluation.

Fig. 12.2 Pollutant concentration at the various distances (0, 100, 200 m) from the injection point, in the simple example considered

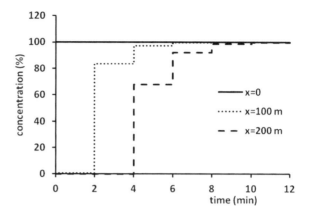

12.7   Crank-Nicolson Numerical Scheme (CTCS)

Instead of the approach described in Sect. 12.6, other forms of discretisation are available in the scientific literature and are currently applied for the water quality modelling. A very interesting form is the implicit Crank-Nicolson numerical scheme, in which the six-point grid is simplified as illustrated in Fig. 12.3.

The time derivative remains at the points x, t and x, $t + 1$, while the space derivatives (first and second order) are moved to an intermediate point x, $t + \frac{1}{2}$. Consequently, the approach becomes *centred time/centred space* (CTCS).

In case of a steady uniform flow in a stream

$$U_x^t = U = \text{constant} \tag{12.19}$$

the derivative of velocity in time and space is zero

$$\frac{\partial U}{\partial t} = 0 \quad \frac{\partial U}{\partial x} = 0 \tag{12.20}$$

and the fundamental transport equation in one-dimensional (1D) form (presented in previous chapters) becomes

$$\frac{\partial C}{\partial t} + \frac{\partial UC}{\partial x} = E\frac{\partial^2 C}{\partial x^2} - kC \Rightarrow \frac{\partial C}{\partial t} + C\frac{\partial U}{\partial x} + U\frac{\partial C}{\partial x} = E\frac{\partial^2 C}{\partial x^2} - kC$$

$$\Rightarrow \frac{\partial C}{\partial t} + U\frac{\partial C}{\partial x} = E\frac{\partial^2 C}{\partial x^2} - kC \tag{12.21}$$

To transform the above differential equation into a discrete expression, the time derivative becomes

$$\frac{\partial C}{\partial t} \rightarrow \frac{C_x^{t+1} - C_x^t}{\Delta t} \tag{12.22}$$

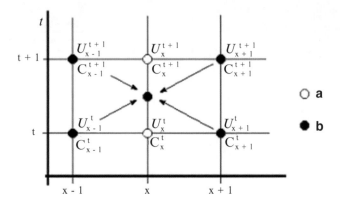

Fig. 12.3 The simplified grid of the Crank-Nicolson scheme: points **a** contain the time derivative and the second space derivative, while points **b** contain both first and second space derivatives

The first space derivative at the intermediate point can be expressed as the arithmetic mean of the derivative at the original points of the grid, also transformed into finite differences

$$\frac{\partial C}{\partial x} \rightarrow \frac{\dfrac{C_{x+1}^{t+1} - C_{x-1}^{t+1}}{2\Delta x} + \dfrac{C_{x+1}^{t} - C_{x-1}^{t}}{2\Delta x}}{2} \tag{12.23}$$

In a similar way, the second derivative, at the same intermediate point, is the arithmetic mean of the second derivative at the same original points

$$\frac{\partial^2 C}{\partial x^2} \rightarrow \frac{\dfrac{C_{x+1}^{t} - 2C_{x}^{t} + C_{x-1}^{t}}{\Delta x^2} + \dfrac{C_{x+1}^{t+1} - 2C_{x}^{t+1} + C_{x-1}^{t+1}}{\Delta x^2}}{2} \tag{12.24}$$

Finally, the reactive term is expressed as

$$k\left(\frac{C_{x}^{t} + C_{x}^{t+1}}{2}\right) \tag{12.25}$$

Consequently, the fundamental differential equation is replaced by

$$\frac{C_{x}^{t} - C_{x}^{t+1}}{\Delta t} - \frac{U}{4\Delta x}\left(C_{x+1}^{t+1} - C_{x-1}^{t+1} + C_{x+1}^{t} - C_{x-1}^{t}\right)$$
$$+ \frac{E}{2\Delta x^2}\left[C_{x+1}^{t} + C_{x-1}^{t+1} - 2\left(C_{x}^{t} + C_{x}^{t+1}\right) + C_{x-1}^{t} + C_{x-1}^{t+1}\right] - \frac{k}{2}\left(C_{x}^{t} + C_{x}^{t+1}\right) = 0 \tag{12.26}$$

The solution of (12.26), based on appropriate initial and downstream boundary conditions, is transformed to the solution of a linear system of equations, for the various steps into which the water body can be discretised, at a given time step.

In the scientific literature, the Crank-Nicolson scheme is considered more reliable than the method previously illustrated. Nevertheless, the persistence of the six-point grid can bring some complexities in the development of the various steps necessary to achieve the solution. The Crank-Nicolson scheme is an implicit scheme; therefore, there is no need for satisfying any stability criterion. Despite of this fact, it is generally recommended that

$$\frac{U\Delta x}{E} \le 2 \qquad (12.27)$$

in order to avoid spatial oscillations (Fletcher 1990).

12.8   The BTCS Numerical Scheme

A further simplification, which is also popular and has given satisfactory results, consists of locating the discrete terms only on four points of the grid, as illustrated in Fig. 12.4. This implicit numerical scheme, known as *backward time/centred space* (BTCS), is often applied in practice and is the basis of the QUAL2 model developed in the United States (EPA 1987). A short description of QUAL2 is included in the appendix of this book.

Assuming, for simplicity, a uniform stream with $U = $ const. as above, the discretisation proceeds as described in the following lines:

$$\frac{\partial C}{\partial t} \rightarrow \frac{C_x^{t+1} - C_x^t}{\Delta t}$$

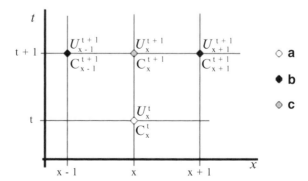

Fig. 12.4 A first case of four-point scheme: BTCS. Point **a** concerns only time, points **b** are related to space and point **c** concerns both time and space

$$\frac{\partial C}{\partial x} \rightarrow \frac{C_{x+1}^{t+1} - C_{x-1}^{t+1}}{2\Delta x}$$

$$\frac{\partial^2 C}{\partial x^2} \rightarrow \frac{C_{x+1}^{t+1} - 2C_x^{t+1} + C_{x-1}^{t+1}}{\Delta x^2} \tag{12.28}$$

while the reactive term is simply referred to the point x, $t + 1$.

The fundamental differential equation is, therefore,

$$\frac{C_x^{t+1} - C_x^t}{\Delta t} + \frac{U}{2\Delta x}\left(C_{x+1}^{t+1} - C_{x-1}^{t+1}\right)$$
$$- \frac{E}{\Delta x^2}\left(C_{x+1}^{t+1} - 2C_x^{t+1} + C_{x-1}^{t+1}\right) + kC_x^{t+1} = 0 \tag{12.29}$$

which after expansion can be written as

$$-\frac{C_x^t}{\Delta t} + \alpha_1 C_x^{t+1} + \alpha_2 C_{x+1}^{t+1} + \alpha_3 C_{x-1}^{t+1} = 0 \tag{12.30}$$

in which

$$\alpha_1 = \frac{1}{\Delta t} + \frac{2E}{\Delta x^2} + k$$
$$\alpha_2 = \frac{U}{2\Delta x} - \frac{E}{\Delta x^2} \tag{12.31}$$
$$\alpha_3 = -\frac{U}{2\Delta x} - \frac{E}{\Delta x^2}$$

Also, this equation, with proper initial and downstream boundary conditions, applied to a sequence of river longitudinal steps at a certain time, gives rise to a linear system of equations, which leads to the solution of the problem. The procedure is illustrated by means of the following example.

As also presented in Chap. 9, a constant pollutant concentration is injected in a cross section of the one-dimensional stream illustrated in Fig. 12.5. The problem consists of finding the pollutant concentration at downstream cross sections at subsequent times.

The water velocity, constant all over the stream, is $U = 0.3$ m/s; the dispersion coefficient, also constant, is $E = 5.0$ m^2/s and the pollutant reaction coefficient $k = 0.00005$ s^{-1}. The problem considers a space discretisation $\Delta x = 25$ m and time steps $\Delta t = 20$ s. Using these values, the coefficients of Eq. (12.30) are evaluated.

$$\alpha_1 = \quad 0.066$$
$$\alpha_2 = \quad 0.004$$
$$\alpha_3 = -0.020$$
$$\alpha_4 = \quad 0.050$$

Fig. 12.5 The one-dimension stream considered for the application of the BTCS scheme

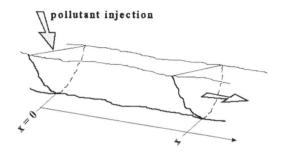

It should be stressed that the BTCS scheme is implicit, and hence it is unconditionally stable.

Adopting the same notations already introduced in preceding chapters, for the first time step starting at $t = 0$, the recursive equation becomes

$$-0.050\ C(x, 0) + 0.066\ C(x, 1) + 0.004\ C(x+1, 1) - 0.020\ C(x-1, 1) = 0$$

Following the adopted assumptions, the equation must be solved for any fixed time interval, for the various spatial steps $x = 1$, $x = 2$, etc. Therefore, for the four spatial steps taken into consideration, the equation gives rise to the following linear system:

$$
\begin{array}{rllllllll}
0.066 & C(1, 1) & +0.004 & C(2, 1) & & & & & = & 2.0 \\
-0.020 & C(1, 1) & +0.066 & C(2, 1) & +0.004 & C(3, 1) & & & = & 0.0 \\
& & -0.020 & C(2, 1) & +0.066 & C(3, 1) & +0.004 & C(4, 1) & = & 0.0 \\
& & & & -0.020 & C(3, 1) & +0.070 & C(4, 1) & = & 0.0
\end{array}
$$

The downstream boundary condition has been posed in accordance with the Neumann rule, as indicated in Sect. 12.5, which can be written in the form

$$C(5, 1) = C(4, 1)$$

The solution of this system is

$$
\begin{array}{ll}
C(1, 1) = 29.74 & C(2, 1) = 8.85 \\
C(3, 1) = 2.63 & C(4, 1) = 0.78
\end{array}
$$

For other time steps, $t = 1, t = 2$, etc., the procedure is similar. In this example, 23 time steps are considered, and the relevant solutions are summarised in Table 12.4. After replacing the time and space steps with the proper terms of the problem

$$
\begin{array}{l}
t = 1 \rightarrow \Delta t = 20\ \text{s} \\
t = 2 \rightarrow 2\Delta t = 40\ \text{s} \\
t = 3 \rightarrow 3\Delta t = 1\ \text{min}
\end{array}
$$

Table 12.4 The solution of the linear system for the various time steps

$t=0$		$t=5$		$t=10$		$t=15$		$t=20$	
$C(0,0)$	100.00	$C(0,5)$	100.00	$C(0,10)$	100.00	$C(0,15)$	100.00	$C(0,20)$	100.00
$C(1,0)$	0.00	$C(1,5)$	85.71	$C(1,10)$	98.83	$C(1,15)$	99.86	$C(1,20)$	99.81
$C(2,0)$	0.00	$C(2,5)$	60.86	$C(2,10)$	92.66	$C(2,15)$	99.05	$C(2,20)$	99.63
$C(3,0)$	0.00	$C(3,5)$	37.43	$C(3,10)$	80.23	$C(3,15)$	96.04	$C(3,20)$	99.14
$C(4,0)$	0.00	$C(4,5)$	20.69	$C(4,10)$	63.70	$C(4,15)$	89.41	$C(4,20)$	97.50
$C(5,0)$	0.00	$C(5,5)$	10.57	$C(5,10)$	46.80	$C(5,15)$	79.57	$C(5,20)$	94.45
$C(6,0)$	0.00	$C(6,5)$	4.81	$C(6,10)$	30.25	$C(6,15)$	63.94	$C(6,20)$	86.14
$t=1$		$t=6$		$t=11$		$t=16$		$t=21$	
$C(0,1)$	100.00	$C(0,6)$	100.00	$C(0,11)$	100.00	$C(0,16)$	100.00	$C(0,21)$	100.00
$C(1,1)$	29.74	$C(1,6)$	90.88	$C(1,11)$	99.34	$C(1,16)$	99.86	$C(1,21)$	99.80
$C(2,1)$	8.85	$C(2,6)$	70.69	$C(2,11)$	95.06	$C(2,16)$	99.33	$C(2,21)$	99.63
$C(3,1)$	2.63	$C(3,6)$	47.97	$C(3,11)$	85.24	$C(3,16)$	97.21	$C(3,21)$	99.28
$C(4,1)$	0.78	$C(4,6)$	29.20	$C(4,11)$	70.72	$C(4,16)$	92.04	$C(4,21)$	98.07
$C(5,1)$	0.23	$C(5,6)$	16.35	$C(5,11)$	54.59	$C(5,16)$	83.89	$C(5,21)$	95.81
$C(6,1)$	0.07	$C(6,6)$	8.10	$C(6,11)$	37.18	$C(6,16)$	69.59	$C(6,21)$	88.84
$t=2$		$t=7$		$t=12$		$t=17$		$t=22$	
$C(0,2)$	100.00	$C(0,7)$	100.00	$C(0,12)$	100.00	$C(0,17)$	100.00	$C(0,22)$	100.00
$C(1,2)$	51.48	$C(1,7)$	94.32	$C(1,12)$	99.62	$C(1,17)$	99.85	$C(1,22)$	99.80
$C(2,2)$	21.78	$C(2,7)$	78.57	$C(2,12)$	96.72	$C(2,17)$	99.49	$C(2,22)$	99.62
$C(3,2)$	8.40	$C(3,7)$	57.79	$C(3,12)$	89.16	$C(3,17)$	98.02	$C(3,22)$	99.36
$C(4,2)$	3.07	$C(4,7)$	38.20	$C(4,12)$	76.79	$C(4,17)$	94.06	$C(4,22)$	98.46
$C(5,2)$	1.08	$C(5,7)$	23.19	$C(5,12)$	61.90	$C(5,17)$	87.46	$C(5,22)$	96.83
$C(6,2)$	0.36	$C(6,7)$	12.40	$C(6,12)$	44.21	$C(6,17)$	74.64	$C(6,22)$	91.06
$t=3$		$t=8$		$t=13$		$t=18$		$t=23$	
$C(0,3)$	100.00	$C(0,8)$	100.00	$C(0,13)$	100.00	$C(0,18)$	100.00	$C(0,23)$	100.00
$C(1,3)$	67.08	$C(1,8)$	96.55	$C(1,13)$	99.77	$C(1,18)$	99.83	$C(1,23)$	99.80
$C(2,3)$	35.78	$C(2,8)$	84.68	$C(2,13)$	97.85	$C(2,18)$	99.57	$C(2,23)$	99.61
$C(3,3)$	16.76	$C(3,8)$	66.53	$C(3,13)$	92.17	$C(3,18)$	98.56	$C(3,23)$	99.40
$C(4,3)$	7.22	$C(4,8)$	47.20	$C(4,13)$	81.88	$C(4,18)$	95.57	$C(4,23)$	98.72
$C(5,3)$	2.94	$C(5,8)$	30.78	$C(5,13)$	68.56	$C(5,18)$	90.36	$C(5,23)$	97.57
$C(6,3)$	1.09	$C(6,8)$	17.64	$C(6,13)$	51.13	$C(6,18)$	79.08	$C(6,23)$	92.85
$t=4$		$t=9$		$t=14$		$t=19$			
$C(0,4)$	100.00	$C(0,9)$	100.00	$C(0,14)$	100.00	$C(0,19)$	100.00		
$C(1,4)$	78.09	$C(1,9)$	97.96	$C(1,14)$	99.83	$C(1,19)$	99.82		
$C(2,4)$	49.11	$C(2,9)$	89.29	$C(2,14)$	98.58	$C(2,19)$	99.61		
$C(3,4)$	26.76	$C(3,9)$	74.02	$C(3,14)$	94.41	$C(3,19)$	98.92		
$C(4,4)$	13.20	$C(4,9)$	55.79	$C(4,14)$	86.06	$C(4,19)$	96.69		
$C(5,4)$	6.07	$C(5,9)$	38.76	$C(5,14)$	74.46	$C(5,19)$	92.66		
$C(6,4)$	2.51	$C(6,9)$	23.66	$C(6,14)$	57.75	$C(6,19)$	82.90		

and respectively

$$x = 1 \rightarrow \quad \Delta x = 25 \text{ m}$$

$$x = 2 \rightarrow \quad 2\Delta x = 50 \text{ m}$$

As it can be noticed, in this discretisation scheme, the space derivatives are referred at an advanced time step, $t + 1$, which then contains some unknowns. In this case, it is necessary to solve an algebraic system with many equations as mentioned before using various methods.

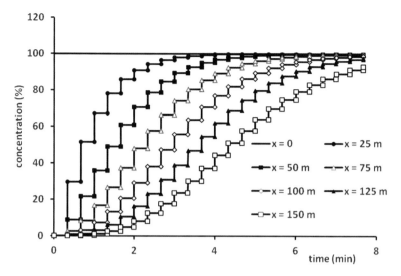

Fig. 12.6 Calculated pollution concentration in the stream by means of the BTCS approach

The final result is shown in Fig. 12.6. It is worthy to notice that the concentration tends asymptotically to values that, at the various locations downstream of the pollutant injection, are progressively lesser than the saturation. This is due to the reaction, formally expressed by coefficient k. It may be of interest to compare Fig. 12.6 with Figs. 9.2 and 9.3 of Chap. 9.

It is also interesting to note that as soon as the distance from the injection point increases, the initial "front" tends to be smoother.

12.9   Explicit Numerical Schemes

Several other four-point schemes are also available, such as the one based on the grid of Fig. 12.7. In this case, the velocity and concentration values at time t are supposed to be known.

The explicit numerical scheme of Fig. 12.7 is *forward time/centred space* (FTCS) and can be developed with the following discrete terms:

$$\frac{\partial C}{\partial t} \rightarrow \frac{C_x^{t+1} - C_x^t}{\Delta t}$$

$$\frac{\partial C}{\partial x} \rightarrow \frac{C_{x+1}^t - C_{x-1}^t}{2\Delta x}$$

$$\frac{\partial^2 C}{\partial x^2} \rightarrow \frac{C_{x+1}^t - 2C_x^t + C_{x-1}^t}{\Delta x^2} \tag{12.32}$$

Fig. 12.7 Another case of four-point scheme: FTCS. Like in the previous case, point **a** concerns only time, points **b** are related to space and point **c** concerns both time and space

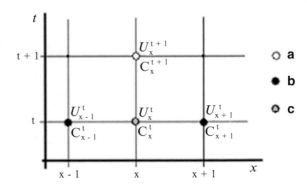

according to which the fundamental differential equation in one-dimensional form becomes

$$\frac{C_x^t - C_x^{t+1}}{\Delta t} + \frac{U}{2\Delta x}\left(C_{x+1}^t - C_{x-1}^t\right) - \frac{E}{\Delta x^2}\left(C_{x+1}^t - 2C_x^t + C_{x-1}^t\right) + kC_x^t = 0 \qquad (12.33)$$

The Eq. (12.36) can be written as

$$C_x^{t+1} = \beta_1 C_x^t + \beta_2 C_{x+1}^t + \beta_3 C_{x-1}^t \qquad (12.34)$$

where

$$\beta_1 = \Delta t\left(\frac{1}{\Delta t} - \frac{2E}{\Delta x^2} - k\right)$$

$$\beta_2 = \Delta t\left(-\frac{U}{2\Delta x} + \frac{E}{\Delta x^2}\right) \qquad (12.35)$$

$$\beta_3 = \Delta t\left(\frac{U}{2\Delta x} + \frac{E}{\Delta x^2}\right)$$

The above equation is for every space and time step a linear equation with only one variable. In this case, the procedure is illustrated by means of the following example, applied to the same stream considered in the previous case.

As mentioned earlier, FTCS is an explicit scheme, and hence it is stable under certain conditions. Specifically for FTCS according to Fletcher (1990),

$$0 \le \left(\frac{U\Delta t}{\Delta x}\right)^2 \le \frac{2E\Delta t}{\Delta x^2} \le 1 \qquad (12.36)$$

Fletcher also recommended for achieving sufficient accuracy that

$$\frac{U\Delta x}{E} \ll \frac{2\Delta x}{U\Delta t} \tag{12.37}$$

Based on the original data, the coefficients of Eq. (12.36) are

$$\beta_1 = 0.679$$
$$\beta_2 = 0.040$$
$$\beta_3 = 0.280$$

For the first time interval $t = 0$ and the space step $x = 1$, the equation gives

$$C(1,1) = 0.679\, C(1,0) + 0.04\, C(2,0) + 0.28\, C(0,0)$$

and for

$$C(1,0) = 0$$
$$C(2,0) = 0$$
$$C(0,0) = 100$$

the solution is

$$C(1,1) = 28.00$$

Proceeding in the same way for the other time and space steps, the solutions obtained are shown in Table 12.5.

Also Fig. 12.8 shows the graphical presentation of the solution.

There are several other explicit numerical procedures, among which the upwind scheme and the Lax-Wendroff scheme are worth mentioning.

According to the well-known upwind numerical scheme, the discretisation of transport equation in one-dimensional form becomes

$$\frac{C_x^{t+1} - C_x^t}{\Delta t} + \frac{U}{\Delta x}\left(C_x^t - C_{x-1}^t\right) - \frac{E}{\Delta x^2}\left(C_{x+1}^t - 2C_x^t + C_{x-1}^t\right) + kC_x^t = 0 \tag{12.38}$$

As proposed by Fletcher (1990), the stability criterion is

$$\frac{U\Delta t}{\Delta x} + \frac{2E\Delta t}{\Delta x^2} \le 1 \tag{12.39}$$

and for achieving accuracy, it is necessary that

$$\frac{U\Delta t}{\Delta x} \ll \frac{2\Delta x}{\Delta x - U\Delta t} \tag{12.40}$$

Table 12.5 Pollution concentration at the various time and space by the FTCS approach

$t = 0$		$t = 3$		$t = 6$		$t = 9$		$t = 12$	
$C(0,1)$	100.00	$C(0,4)$	100.00	$C(0,7)$	100.00	$C(0,10)$	100.00	$C(0,13)$	100.00
$C(1,1)$	28.00	$C(1,4)$	69.64	$C(1,7)$	85.30	$C(1,10)$	91.18	$C(1,13)$	94.10
$C(2,1)$	0.00	$C(2,4)$	29.51	$C(2,7)$	56.86	$C(2,10)$	74.04	$C(2,13)$	84.16
$C(3,1)$	0.00	$C(3,4)$	6.67	$C(3,7)$	27.45	$C(3,10)$	48.73	$C(3,13)$	65.21
$C(4,1)$	0.00	$C(4,4)$	0.61	$C(4,7)$	9.07	$C(4,10)$	25.09	$C(4,13)$	42.70
$C(5,1)$	0.00	$C(5,4)$	0.00	$C(5,7)$	0.43	$C(5,10)$	3.83	$C(5,13)$	22.66
$C(6,1)$	0.00	$C(6,4)$	0.00	$C(6,7)$	0.00	$C(6,10)$	2.11	$C(6,13)$	8.75
$C(7,1)$	0.00	$C(7,4)$	0.00	$C(7,7)$	0.00	$C(7,10)$	0.25	$C(7,13)$	1.53
$C(8,1)$	0.00	$C(8,4)$	0.00	$C(8,7)$	0.00	$C(8,10)$	0.00	$C(8,13)$	0.35
$t = 1$		$t = 4$		$t = 7$		$t = 10$		$t = 13$	
$C(0,2)$	100.00	$C(0,5)$	100.00	$C(0,8)$	100.00	$C(0,11)$	100.00	$C(0,14)$	100.00
$C(1,2)$	47.01	$C(1,5)$	76.47	$C(1,8)$	88.20	$C(1,11)$	92.18	$C(1,14)$	94.93
$C(2,2)$	7.84	$C(2,5)$	39.80	$C(2,8)$	63.59	$C(2,11)$	78.03	$C(2,14)$	86.50
$C(3,2)$	0.00	$C(3,5)$	12.81	$C(3,8)$	34.92	$C(3,11)$	54.82	$C(3,14)$	69.55
$C(4,2)$	0.00	$C(4,5)$	2.28	$C(4,8)$	13.84	$C(4,11)$	31.05	$C(4,14)$	48.16
$C(5,2)$	0.00	$C(5,5)$	0.00	$C(5,8)$	2.54	$C(5,11)$	6.33	$C(5,14)$	27.69
$C(6,2)$	0.00	$C(6,5)$	0.00	$C(6,8)$	0.12	$C(6,11)$	1.07	$C(6,14)$	12.35
$C(7,2)$	0.00	$C(7,5)$	0.00	$C(7,8)$	0.00	$C(7,11)$	0.76	$C(7,14)$	3.50
$C(8,2)$	0.00	$C(8,5)$	0.00	$C(8,8)$	0.00	$C(8,11)$	0.08	$C(8,14)$	0.43
$t = 2$		$t = 5$		$t = 8$		$t = 11$		$t = 14$	
$C(0,3)$	100.00	$C(0,6)$	100.00	$C(0,9)$	100.00	$C(0,12)$	100.00	$C(0,15)$	100.00
$C(1,3)$	60.23	$C(1,6)$	81.51	$C(1,9)$	90.43	$C(1,12)$	93.18	$C(1,15)$	96.79
$C(2,3)$	18.49	$C(2,6)$	48.95	$C(2,9)$	69.27	$C(2,12)$	81.37	$C(2,15)$	88.46
$C(3,3)$	2.20	$C(3,6)$	19.94	$C(3,9)$	42.07	$C(3,12)$	60.31	$C(3,15)$	73.37
$C(4,3)$	0.00	$C(4,6)$	5.14	$C(4,9)$	19.28	$C(4,12)$	36.96	$C(4,15)$	53.28
$C(5,3)$	0.00	$C(5,6)$	0.64	$C(5,9)$	1.72	$C(5,12)$	17.81	$C(5,15)$	32.78
$C(6,3)$	0.00	$C(6,6)$	0.00	$C(6,9)$	0.79	$C(6,12)$	5.47	$C(6,15)$	16.28
$C(7,3)$	0.00	$C(7,6)$	0.00	$C(7,9)$	0.03	$C(7,12)$	1.24	$C(7,15)$	5.85
$C(8,3)$	0.00	$C(8,6)$	0.00	$C(8,9)$	0.00	$C(8,12)$	0.26	$C(8,15)$	1.27

Similarly, with the Lax-Wendroff numerical scheme, the discretisation of transport equation in one-dimensional form (Lax and Wendroff 1964) is

$$\frac{C_x^{t+1} - C_x^t}{\Delta t} + \frac{U}{2\Delta x}\left(C_{x+1}^t - C_{x-1}^t\right) - \frac{(E + 0.5U^2\Delta t)}{\Delta x^2}\left(C_{x+1}^t - 2C_x^t + C_{x-1}^t\right) + kC_x^t = 0$$

(12.41)

and the relevant stability criterion is (Fletcher 1990)

$$0 \le \left(\frac{U\Delta t}{\Delta x}\right)^2 \le \frac{2E\Delta t + (U\Delta t)^2}{\Delta x^2} \le 1$$

(12.42)

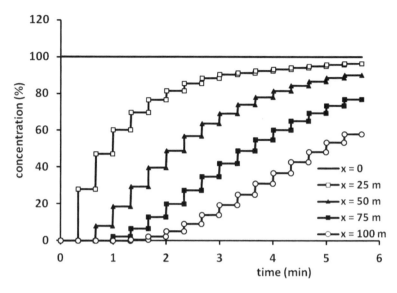

Fig. 12.8 Calculated pollution concentration in the stream by means of the FTCS approach

while it is also recommended that the following condition should be met in order to avoid spatial oscillations:

$$\frac{U\Delta x}{E} \leq 2 \tag{12.43}$$

The errors in numerical simulations which are caused from the first-order numerical schemes are diffusion errors and from the second-order numerical schemes are dispersion errors, respectively. Generally, the parabolic part of the transport equation is a diffusion term. If the first-order truncation error in the discretisation becomes significant in comparison to the diffusion term in the transport equation, accuracy problems of the numerical solution arise.

The Lax-Wendroff is a numerical scheme with second-order accuracy for the hyperbolic part of the transport equation in comparison with the FTCS or the upwind numerical schemes which have first-order accuracy. This is the reason that for the Lax-Wendroff numerical scheme, there is no accuracy requirement but only a recommendation for avoiding spatial oscillations (dispersion errors).

Two numerical examples with the Lax-Wendroff numerical scheme are now presented. The first one consists of the 1D transport simulation of a conservative pollutant ($k = 0$ s^{-1}) with no other injection or subtraction in a river reach, which has a length of 8,000 m. The dispersion coefficient is $E = 70$ m^2/s, and the flow velocity $U = 1.5$ m/s is constant in the entire computational field.

Concerning the boundary conditions, the initial pollutant concentration is zero in the entire field, the upstream boundary concentration is $C_0 = 150$ g/m^3 (Dirichlet) and, at the downstream boundary, the Neumann condition, already defined in Eq. (12.17), is assumed. Simulation results are presented in Fig. 12.9.

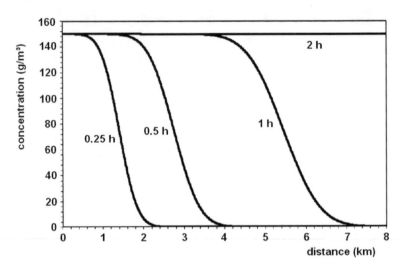

Fig. 12.9 Concentration of conservative pollutant at selected time horizons

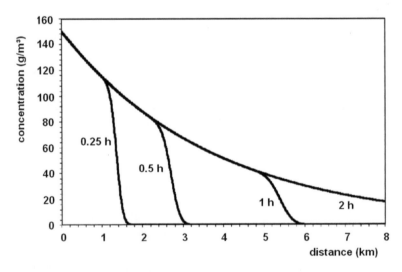

Fig. 12.10 Concentration of nonconservative pollutant at selected time horizons

Similarly, the simulation results for 1D transport of a nonconservative pollutant with $k = 0.0004$ s^{-1} and a dispersion coefficient $E = 8$ m^2/s are presented in Fig. 12.10.

In both figures, the concentration is expressed by means of proper units in the metric system, and the graphs are in the form of continuous lines.

12.10   Final Comments on the FDM

The preceding paragraphs describe few of the most common ways of applying the FDM to solve typical problems of river and stream pollution. Because each method starts from a particular procedure for the transformation of a continuous reality into discrete terms, it is reasonable to expect different solutions. In fact, Fig. 12.11 shows a comparison among various solutions for some significant cross sections of the river, downstream of the point of pollutant injection. In the same figure, also the corresponding analytical solution is shown.

The considered procedures are the Crank-Nicolson method, the BTCS and the FTCS numerical schemes.

The discrepancies shown in the figure are relatively small and are expected to be of the same order of magnitude with the measured values that can be observed in a river, taking into account the real complexity of a practical measuring procedure.

It is expected that adopting smaller time and space intervals, the discrepancies will vanish, and the final result will be closer to the analytical solution. Obviously, decreasing the size of time and space intervals inevitably increases the computational burden. A reasonable compromise is therefore recommended in the discretisation process.

The advantages of the explicit schemes are the less complex algorithms and the faster implementation of iterations. On the other hand, although the procedure of iterations in the algorithms based on implicit numerical schemes is slower, the

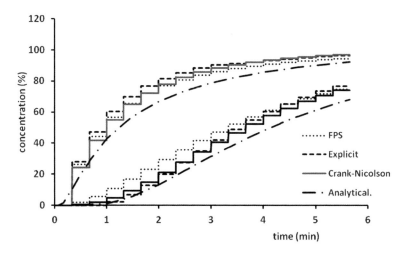

Fig. 12.11 Comparison of different solutions for a problem of continuous injection of infinite duration, adopting the analytical and several numerical procedures. The problem data are as follows: $U = 0.3$ m/s. $E = 5.0$ m^2/s, $k = 0.00005$ s^{-1}, $\Delta x = 25$ m and $\Delta t = 20$ s. The values refer to two different cross sections downstream from the pollutant injection ($x = 25$ m and $x = 75$ m)

non-existence of stability criterion allows big values for time steps and hence less computational time.

Generally, the FDM requires a grid of constant and uniform sizes, all over the computational field. However, its main drawback is the requirement of a rectangular computational field. This can increase the number of steps and require longer computational time, because all parts of the water body should be treated in the same way, even those that require less attention. Other numerical methods, such as the FEM or the FVM and other methods of grid construction, allow for certain improvement (Szymkiewicz 2010), as it will be explained in the next chapter.

Despite this fact, the FDM is by far the simplest numerical method for developing algorithms, especially in 1D models; is faster in several cases in comparison with other methods; and is useful and easy to use in practical problems.

References

Chapra SC (1997) Surface water-quality modeling. WCB-McGraw Hill, Boston

Eheart JW (2006) Some numerical properties of explicit solutions to the one-dimensional conservation equation. Water Int 31(2):252–258

EPA (U.S. Environmental Protection Agency) (1987) The enhanced stream water quality models QUAL2E and QUAL2E-UNCAS. EPA/600/3-87/007, Environmental Research Laboratory, Athens, GA

Fernandez C, Karney B (2001) Numerical solution of the advection-dispersion-reaction equation under transient hydraulic condition. In: Proceedings of the 29th IAHR congress, Beijing, China, theme B, pp 58–72

Fletcher CAJ (1990) Computational techniques for fluid dynamics 1 – Fundamental and general techniques, 2nd edn. Springer, Berlin, p 401

Lax PD, Wendroff B (1964) Difference schemes for hyperbolic equations with high order of accuracy. Commun Pure Appl Math 17:381–398

Stamou AI (1991) Numerical modelling of water quality in rivers using quick scheme. In: Tsakiris G (ed) Proceedings of the European conference advances in water resources technology, Athens, Greece, Balkema Publishers, Rotterdam, The Netherland, pp 397–405

Szymkiewicz R (2010). Numerical modeling in open channel hydraulics. Water sciences and technology library, vol 83. Springer, Dordrecht, The Netherlands

Zhihua O, Elsworth D (1989) An adaptive characteristics method for advective-diffusive transport. Appl Math Model 13:682–692

Chapter 13
The Finite Element Method

Abstract Apart from the finite difference method, the numerical calculus has developed other numerical methods for the integration of the differential equations, with the help of the advanced computing devices. The finite element method (FEM) is one of the most promising methods, which has been successfully applied to many engineering problems. For the water quality models, despite of some difficulties during the first steps of its application, the method gives good opportunities for stable and reliable solutions, especially when irregular river geometry requires a representation by discrete terms of various shapes and sizes. In this chapter, the FEM is applied to water quality problems of one dimension.

13.1 Fundamental Aspects

The progress in computer technology has favoured the progress in numerical methods, creating tools and procedures that have allowed the finite element method to be greatly improved and applicable in many practical engineering problems. Originally developed for mechanical structures, the method has been adopted by the water scientists and engineers to solve water- and pollution-related problems. Up to now, apart from the mathematical interpretation of some physical aspects, it can provide interesting solutions in water quality modelling (Casulli and Cheng 1992; Gelinas and Doss 1981; Lee and Seo 2007).

The following paragraphs present the fundamentals of the method, in order to show how it can be applied to river pollution problems, putting also into evidence its peculiarities in comparison with the other methods already described. More details and comprehensive descriptions are available in the abundant scientific literature on the subject (Evans 2000; Li and Chen 1989; Strang and Fix 1973; Zienkiewicz 1972), while recent scientific contributions have refined several aspects, improving, in particular, the calculation procedures.

In the application to water-related problems, the finite element theory has been developed for the three-dimensional approach, with algorithms depending on the

M. Benedini and G. Tsakiris, *Water Quality Modelling for Rivers and Streams*,
Water Science and Technology Library 70, DOI 10.1007/978-94-007-5509-3_13,
© Springer Science+Business Media Dordrecht 2013

three space variables and time. Also two-dimensional analysis has been extensively used. For the sake of a clear presentation of the fundamental aspects, the following pages are restricted to the one-dimensional approach. This is not a penalising restriction, because, as already seen in the preceding chapters, the one-dimensional approach can give a response sufficiently reliable in the most common problems of river pollution, simulating satisfactorily pollutant behaviour.

As in the numerical procedures already presented, the crucial point of the FEM is the risk of instability due to the simultaneous presence of parabolic and hyperbolic terms in the fundamental equation. Several ways of proceeding have been devised, the majority of which are based on the possibility to interpret separately the advection and the dispersion pollutant transport (Celia et al. 1990; Al-Lavatia et al. 1999).

13.2   The Basic Algorithm for a Continuous Pollutant Injection

For the one-dimensional approach with a continuous polluting source of infinite duration, the fundamental partial differential equation (PDE) is written as

$$\frac{\partial C}{\partial t} = E\frac{\partial^2 C}{\partial x^2} - U\frac{\partial C}{\partial x} - kC \pm S \tag{13.1}$$

Equation (13.1) must comply with the initial and boundary conditions written as follows:

$$\begin{cases} C(x,0) = 0 \\ C(0,t) = \gamma \\ \dfrac{\partial C(L,0)}{\partial x} = 0 \end{cases} \quad \text{for} \quad 0 \le x \le L \tag{13.2}$$

in the time interval $0 \le t \le T$, having set $\gamma = C(0,0)$ for the concentration at the injection cross section $(x = 0)$ at $t = 0$. L is the length of the river stretch studied.

Introducing the function

$$\Psi(x,t) = C(x,t) - \gamma \tag{13.3}$$

Equation (13.1) becomes

$$\frac{\partial \Psi(x,t)}{\partial t} = E\frac{\partial^2 \Psi(x,t)}{\partial x^2} - U\frac{\partial \Psi(x,t)}{\partial x} - k(\Psi(x,t) + \gamma) \pm S \tag{13.4}$$

and the initial and boundary conditions (13.2) are equivalent to

$$\begin{cases} \Psi(x,0) = -\gamma \\ \Psi(0,t) = 0 \\ \dfrac{\partial \Psi(L,0)}{\partial x} = 0 \end{cases} \tag{13.5}$$

In this presentation, for the sake of simplicity, the terms U, E, k and γ are constant for all the river stretch examined. It is clear that this is not a restriction to the method, which, in its more refined form, can encompass any kind of time and space variability.

In a quite general approach, there are no restrictions also on the duration of an elementary time interval $\Delta t = t_{n+1} - t_n$; however, for simplicity, Δt is assumed constant for all the time period considered, which is made up by a number of N intervals. The river stretch, of total length L, is also split into M elementary intervals $\Delta x = l$, assumed to be constant.

With the notation

$$\Psi^n = \Psi(x, t_n)$$

being Ψ^n function only of the x coordinate, the left-hand side of (13.4) can be approximated by

$$\frac{\partial \Psi^n}{\partial t} \cong \frac{\Psi^n - \Psi^{n-1}}{\Delta t}$$

For a given value of t, the problem becomes only a function of the x variable and the partial derivatives can be replaced by ordinary derivatives.

For the time steps $n = 1, \ldots, N$, the Eq. (13.4) gives rise to a system identified by the current expression

$$\Delta t E \frac{d^2 \Psi^n}{dx^2} - \Delta t U \frac{d\Psi^n}{dx} - (1 + k\Delta t)\Psi^n = F^{n-1} \tag{13.6}$$

in which

$$F^{n-1} = -\Psi^{n-1} + k\gamma\Delta t \mp S\Delta t \tag{13.7}$$

The problem is now transformed into the search of a rational expression of Ψ^n that satisfies (13.6). According to the well-known Galerkin method (O'Neil 1981), the solution can be approximated by a linear combination

$$\Psi_M^n = \sum_{j=1}^{M} a_j^n \varphi_j(x)$$

of a family of functions $\varphi_j(x)$ and some coefficients a_j^n, whose significance will be explained in the following paragraphs. Concerning the functions $\varphi_j(x)$, it must be added that, in line with the conditions of $\Psi''(x)$ expressed in (13.5), they satisfy the boundary conditions

$$\varphi_j(0) = 0 \qquad\qquad \frac{d\varphi(L)}{dx} = 0$$

for the elementary space intervals $j = 1, 2, 3, ...M$.

It is of particular interest the family of *linear* functions $\varphi_j(x)$, characterised by

$$\varphi_j(x) = \begin{cases} 0 & \text{for} \quad x \leq (j-1)l \\ \frac{x - (j-1)l}{l} & \text{for} \quad (j-1)l \leq x \leq jl \\ 1 - \frac{x - jl}{l} & \text{for} \quad jl \leq x \leq (j+1)l \\ 0 & \text{for} \quad x \geq (j+1) \end{cases} \qquad (13.8)$$

for $j = 1, 2, \ldots M - 1$.

The last elements of the family of the linear functions are defined by

$$\varphi_M(x) = \begin{cases} 0 & \text{for} \quad x \leq x_{(M-1)l} \\ \frac{x - (M-1)l}{l} & \text{for} \quad x_{(M-1)l} \leq x \leq x_{Ml} \\ 1 & \text{for} \quad x_{Ml} \leq x \leq x_{(M+1)l} \end{cases} \qquad (13.9)$$

where

$$l = \frac{L}{M+1}$$

Figure 13.1 is the plot of $\phi_j(x)$ on the x-axis, along the river stretch, following the original considerations about the geometrical interpretation of the river. Particular attention should be paid to the length of Δx in the river stretch, l, which should be significant for the real sizes of the water body. A too short l is not suitable for a river, in which the hydraulic and geometrical terms are usually referred to reaches of the order of 10 m. A too great l could miss some details of the natural river course.

The coefficients a_j^n, introduced in the mathematical procedure, have a physical significance. In fact, according to (13.3), it is easy to show that

$$a_j^n = C(x_j, t_n) - \gamma \qquad (13.10)$$

Their dimension is therefore that of a concentration. They enable one to find the values of pollutant concentration, namely, the final solution of the problem.

The expressions and functions introduced in the preceding lines allow now the Eq. (13.6) to be transformed in a way that permits its projection to the whole

Fig. 13.1 Configuration of the family of finite linear elements along the x-axis

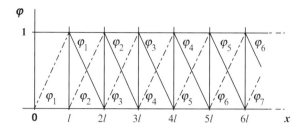

water body. Multiplying by $\phi_j(x)$ both sides and integrating over the $[0,L]$ interval, the following expression can be obtained:

$$\Delta t \int_0^L E \frac{d^2\Psi^n}{dx^2} \varphi_j dx - \Delta t \int_0^L U \frac{d\Psi^n}{dx} \varphi_j dx - (1 + k\Delta t) \int_0^L \Psi^n \varphi_j dx$$

$$= \int_0^L F^{n-1} \varphi_j dx \qquad (13.11)$$

The integrals of (13.11) can be considered separately.

The first integral, proceeding by parts and recalling (13.5), gives

$$\int_0^L E \frac{d^2\Psi^n}{dx^2} \varphi_j dx = -E \int_0^L \frac{d\Psi^n}{dx} \frac{d\varphi_j}{dx} dx$$

Similarly, the second integral, also integrating by parts, gives

$$\int_0^L U \frac{d\Psi^n}{dx} \varphi_j dx = -U \int_0^L \Psi^n \frac{d\varphi_j}{dx} dx$$

Consequently, Eq. (13.11) can be written in the form

$$-\Delta t E \int_0^L \frac{d\Psi^n}{dx} \frac{d\varphi_j}{dx} dx + \Delta t U \int_0^L \Psi^n \frac{d\varphi_j}{dx} dx - (1 + k\Delta t) \int_0^L \Psi^n \varphi_j dx$$

$$= \int_0^L F^{n-1} \varphi_j dx \qquad (13.12)$$

in which Ψ^n and ϕ_j are functions only of x for all the space elements identified by $j = 1, 2, \ldots, M$.

Fig. 13.2 The first finite linear elements

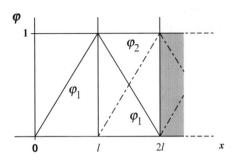

In the computation of (13.12), it is convenient to distinguish the three cases

$$j = 1 \qquad\qquad 2 \le j \le M - 1 \qquad\qquad j = M$$

which correspond, respectively, to Figs. 13.2, 13.4 and 13.5.

From the grid illustrated in Fig. 13.1, it is now possible to consider separately the first two elements, as in Fig. 13.2.

Taking then into account the specifications of the preceding paragraphs, the first integral of (13.12) gives

$$\int_0^L \frac{d\Psi^n}{dx} \frac{d\varphi_1}{dx} dx = \int_0^l \frac{1}{l^2} a_1^n dx - \int_l^{2l} \frac{1}{l^2}(a_1^n - a_2^n) dx = \frac{1}{l}(2a_1^n - a_2^n) \qquad (13.13)$$

being identically

$$\varphi_1 = 0$$

for x outside the interval $[0, 2l]$.

In the same way, the second integral of (13.12) gives

$$\int_0^L \Psi^n \frac{d\varphi_1}{dx} dx = \int_0^l \frac{a_1^n}{l^2} x dx - \int_l^{2l} \frac{1}{l}\left[a_1^n\left(1 - \frac{x-l}{l}\right) + a_2^n\left(\frac{x-l}{l}\right) \right] dx$$

$$= -\frac{1}{2} a_2^n \qquad (13.14)$$

and the third integral

$$\int_0^L \Psi^n \varphi_1 dx = \int_0^l \frac{a_1^n}{l^2} x^2 dx + \int_l^{2l} \left[a_1^n \left(1 - \frac{x-l}{l}\right)^2 + a_2^n \left(\frac{x-l}{l}\right)\left(1 - \frac{x-l}{l}\right) \right] dx$$

$$= \frac{2}{3} la_1^n + \frac{1}{6} la_2^n \tag{13.15}$$

Recalling Eq. (13.7), the known term in the right-hand side, following the same procedure, is transformed into

$$\int_0^L F^{n-1} \varphi_1 dx = -\int_0^L \Psi^{n-1} \varphi_1 dx + k\gamma\Delta t \int_0^L \varphi_1 dx \pm \Delta t \int_0^L \varphi_1 S dx \tag{13.16}$$

and the first integral in its right-hand side, just replacing n with $n-1$ in (13.15), becomes

$$\int_0^L \Psi^{n-1} \varphi_1 dx = \frac{2}{3} la_1^{n-1} + \frac{1}{6} la_2^{n-1}$$

while both the second and third integrals are simply

$$\int_0^L \varphi_1(x) dx = 1$$

Consequently, for the first element of the φ_j family, Eq. (13.12) is transformed into

$$E \frac{\Delta t}{l} \left[2a_1^n - a_2^n \right] + U \frac{\Delta t}{2} a_2^n - (1 + k\Delta t) \left[\frac{2}{3} la_1^n + \frac{l}{6} a_2^n \right]$$

$$= -\frac{2}{3} la_1^{n-1} - \frac{1}{6} la_2^{n-1} + k\gamma l\Delta t \pm Sl\Delta t \tag{13.17}$$

In a similar way, the second element (and any intermediate element identified by $j = 2, 3, \dots M - 1$) can be considered separately as in Fig. 13.3.

Proceeding in the same way as before, the first, second and third integrals of (13.12) give, respectively,

$$\int_0^L \frac{d\Psi^n}{dx} \frac{d\varphi_j}{dx} dx = -\frac{1}{l} \left(-a_{j-1}^n + 2a_j^n + a_{j+1}^n \right)$$

Fig. 13.3 The intermediate
linear finite elements

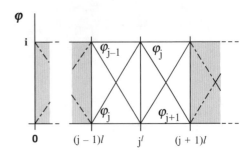

$$\int_0^L \Psi^n \frac{d\varphi_J}{dx} dx = \frac{1}{2} a_{j-1}^n - \frac{1}{2} a_{j+1}^n$$

$$\int_0^L \Psi^n \varphi_j dx = \frac{1}{6} l a_{j-1}^n + \frac{2}{3} l a_j^n + \frac{1}{6} l a_{j+1}^n$$

and the known term in the right-hand side

$$\int_0^L F^{n-1} \varphi_j dx = \frac{1}{6} l a_{j-1}^{n-1} + \frac{2}{3} l a_j^{n-1} + \frac{1}{6} l a_{j+1}^{n-1} - k\gamma l \Delta t \pm l S \Delta t$$

Equation (13.12) becomes, therefore,

$$E \frac{\Delta t}{l} \left[-a_{j-1}^n + 2a_j^n - a_{j+1}^n \right] - U \frac{\Delta t}{2} \left[a_{j-1}^n - a_{j+1}^n \right] - (1 + k\Delta t) \left[\frac{1}{6} l a_{j-1}^n + \frac{2}{3} l a_j^n + \frac{l}{6} l a_{j+1}^n \right]$$

$$= -\frac{1}{6} l a_{j-1}^{n-1} - \frac{2}{3} l a_j^{n-1} - \frac{1}{6} l a_{j+1}^{n-1} + k\gamma l \Delta t \pm S l \Delta t$$

$$(13.18)$$

Finally, for the last element, as in Fig. 13.4,
the three integrals of left-hand side of (13.12) are transformed, respectively, into

$$\int_0^L \frac{d\Psi^n}{dx} \frac{d\varphi_M}{dx} dx = -\frac{1}{l} \left(a_M^n - a_{M-1}^n \right)$$

$$\int_0^L \Psi^n \frac{d\varphi_M}{dx} dx = \frac{1}{2} a_{M-1}^n + \frac{1}{2} a_M^n$$

Fig. 13.4 The last linear
finite element

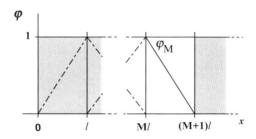

$$\int_0^L \Psi^n \varphi_M \mathrm{d}x = \frac{1}{6} l a^n_{M-1} + \frac{5}{6} l a^n_M$$

and the right-hand side

$$\int_0^L F^{n-1} \varphi_j \mathrm{d}x = \frac{1}{6} l a^{n-1}_{M-1} + \frac{5}{6} l a^{n-1}_M - k\gamma l \Delta t \pm S l \Delta t$$

Equation (13.12) is transformed into

$$E \frac{\Delta t}{l} \left[a^n_M - a^n_{M-1} \right] - U \Delta t \left[\frac{1}{2} a^n_{M-1} + \frac{1}{2} a^n_M \right] - (1 + k\Delta t) \left[\frac{1}{6} l a^n_{M-1} + \frac{8}{6} l a^n_M \right]$$

$$= -\frac{1}{6} l a^{n-1}_{M-1} - \frac{8}{6} l a^{n-1}_M + k\gamma l \Delta t \pm S l \Delta t$$

$$(13.19)$$

At any time step n, the boundary conditions

$$a^n_0 = 0 \qquad\qquad a^n_M = a^n_{M+1}$$

have to be considered.

Equations (13.17), (13.18) and (13.19) can be rearranged for $j = 1$ in the form

$$E \Delta t \frac{1}{l} \left(2 a^n_1 - a^n_2 \right) + U \Delta t \left(\frac{1}{2} a^n_2 \right) - (1 + k\Delta t) l \left(-\frac{2}{3} a^n_1 + \frac{1}{6} a^n_2 \right)$$

$$= -l \left(\frac{2}{3} a^{n-1}_1 + \frac{1}{6} a^{n-1}_2 \right) - k\gamma l \Delta t \pm S l \Delta t \qquad\qquad (13.20a)$$

For $j = 2,\ldots,M - 1$ the following is valid:

$$E\frac{\Delta t}{l}\left[-a_{j-1}^n + 2a_j^n - a_{j+1}^n\right] - U\frac{\Delta t}{2}\left[a_{j-1}^n - a_{j+1}^n\right] - (1 + k\Delta t)\left[\frac{1}{6}la_{j-1}^n + \frac{2}{3}la_j^n + \frac{1}{6}la_{j+1}^n\right]$$

$$= -\frac{1}{6}la_{j-1}^{n-1} - \frac{2}{3}la_j^{n-1} - \frac{1}{6}la_{j+1}^{n-1} + k\gamma l\Delta t \pm Sl\Delta t \tag{13.20b}$$

and for $j = M$,

$$E\Delta t\frac{1}{l}\left(a_M^n - a_{M-1}^n\right) + U\Delta t\left(\frac{1}{2}a_{M-1}^n + \frac{1}{2}a_M^n\right) - (1 + k\Delta t)l\left(\frac{1}{6}a_{M-1}^n + \frac{8}{6}a_M^n\right)$$

$$= -l\left(\frac{1}{6}a_{M-1}^{n-1} + \frac{8}{6}a_M^{n-1}\right) - k\gamma l\Delta t \pm Sl\Delta t \tag{13.20c}$$

As a result a linear system of equations is formulated that, for any value of n, gives

$$a_1^n, a_2^n, \ldots a_M^n$$

and finally, following Eq. (13.10), it gives the values of the pollutant concentration at any element and instant. At any n-step the boundary conditions

$$a_0^n = 0 \qquad\qquad a_{M+1}^n = a_M^n$$

are considered.

The system can be written in vector-matrix form, introducing for the left-hand side the symmetric square matrix $\mathbf{A}$ of order M

$$\mathbf{A} = \begin{vmatrix}
2 & -1 & 0 & \cdots & & & \cdots & 0 \\
-1 & 2 & -1 & 0 & \cdots & & \cdots & 0 \\
0 & -1 & 2 & -1 & 0 & & \cdots & 0 \\
0 & 0 & -1 & 2 & -1 & & \cdots & 0 \\
0 & \cdots & \cdots & \cdots & & & \cdots & 0 \\
0 & \cdots & \cdots & \cdots & \cdots & & \cdots & 0 \\
0 & \cdots & \cdots & \cdots & \cdots & 0 & -1 & 2 & -1 \\
0 & \cdots & \cdots & \cdots & \cdots & & 0 & -1 & 1
\end{vmatrix} \tag{13.21}$$

the symmetric square matrix $\mathbf{B}$

$$\mathbf{B} = \begin{vmatrix} 0 & -1 & 0 & \dots & & & \dots & 0 \\ 1 & 0 & -1 & 0 & \dots & & \dots & 0 \\ 0 & 1 & 0 & -1 & 0 & & \dots & 0 \\ 0 & 0 & 1 & 0 & -1 & & \dots & 0 \\ 0 & \dots & \dots & \dots & \dots & & \dots & 0 \\ 0 & \dots & \dots & \dots & \dots & & \dots & 0 \\ 0 & \dots & \dots & \dots & \dots & 0 & 1 & 0 & -1 \\ 0 & \dots & \dots & \dots & \dots & \dots & 0 & 1 & 1 \end{vmatrix} \qquad (13.22)$$

and the symmetric square matrix $\mathbf{C}$

$$\mathbf{C} = \begin{vmatrix} 4 & 1 & 0 & \dots & \dots & & \dots & 0 \\ 1 & 4 & 1 & 0 & \dots & & \dots & 0 \\ 0 & 1 & 4 & 1 & \dots & & \dots & 0 \\ 0 & 0 & 1 & 4 & 1 & & \dots & 0 \\ 0 & \dots & \dots & \dots & \dots & & \dots & 0 \\ 0 & \dots & \dots & \dots & \dots & & \dots & 0 \\ 0 & \dots & \dots & \dots & \dots & 0 & 1 & 4 & 1 \\ 0 & 0 & \dots & \dots & \dots & \dots & 0 & 1 & 8 \end{vmatrix} \qquad (13.23)$$

as well as the vectors

$$\mathbf{Q}^n = \left[a_1^n, a_2^n, \dots a_M^n \right] \qquad (13.24)$$

and

$$\mathbf{Q}^{n-1} = \left[a_1^{n-1}, a_2^{n-1}, \dots a_M^{n-1} \right] \qquad (13.25)$$

Similarly, for the known terms in the right-hand side, the vector

$$\mathbf{L} = [l, l, \dots 1] \qquad (13.26)$$

can be identified.

The system becomes, therefore, being $\kappa = 1 + k\Delta t$,

$$\left[-\frac{E\Delta t}{l} \mathbf{A} + \frac{U\Delta t}{2} \mathbf{B} - \frac{\kappa l}{6} \mathbf{C} \right] \mathbf{Q}^n = -l\mathbf{C}\mathbf{Q}^{n-1} + l\Delta t (k\gamma \pm S)\mathbf{L} \qquad (13.27)$$

The described algorithms follow the general theory introduced by Galerkin.

System (13.27) can be applied for the problem of pollution transport, but unfortunately, it is conditioned by some well-defined mathematical assumptions, which are not always consistent with the reality of a river. Besides appropriate boundary and initial conditions, the terms of (13.27) have to comply with particular constraints involving the duration of the time intervals and the value of l.

Consequently, the solution can give very often unexpected results, sometimes negative, and unexplainable oscillations.

To overcome such a backlash, the most advanced mathematical techniques provide some amendments.

13.3   The Ritz-Galerkin Approach

The following paragraphs deal with a procedure due to Douglas and Russel (1982), in an advanced development of the differential equations theory. The approach is based on the concept, originally proposed by Ritz, of an "energy functional", to be minimised as a function of the geometrical and physical terms of the problem. A complete description of the procedure requires complex mathematical steps, extensively treated in the scientific literature (Zienkiewicz 1972; Przemienieki 1981; Strang and Fix 1973; Evans 2000). The following paragraphs deal with a simplified explanation, stressing the solution algorithms suitable to be applied in the most common cases of river and stream pollution.

In the one-dimensional approach of a stream, the case of continuous and constant pollutant injection, already proposed in the preceding chapter, gives rise to the linear Eqs. (13.17), (13.18) and (13.19), which comprise a linear system of equations.

A way to overcome the difficulties related to the numerical stability of the computation consists of replacing the second integral of (13.11)

$$\int\limits_0^L \frac{d\Psi^n}{dx}\varphi_j dx$$

by the integral

$$\int\limits_0^L \frac{d\Psi^{n-1}}{dx}\varphi_j dx$$

which, in the consequent interactive scheme, makes up the initial value of each step, instead of its unknown solution. The corresponding constraints for the l and Δt steps, aiming at increasing the numerical stability, are less restrictive than those of the method described in the preceding paragraphs.

The time and longitudinal steps Δt and Δx are chosen so that

$$\Delta x = l = U\Delta t \tag{13.28}$$

Fig. 13.5 The forward scheme for the Ritz-Galerkin method

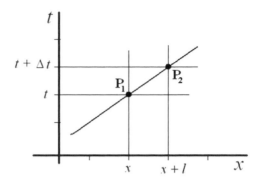

expression which is also the Courant-Friedrichs-Lewy (CFL) condition, already introduced in Chap. 12, restricted to the equality. According to the general theory, it can be proved that the numerical dispersion, characteristic of this kind of mathematical applications, is reduced to a minimum, with a very low probability of unstable solutions. Such a constraint can be more restrictive than those already introduced in the preceding chapters, but can be adopted for the majority of river pollution cases.

Neglecting the details, a method proposed by Douglas and Russel (1982) is described in the next paragraphs. It improves greatly the numerical stability of the integration process and is based on the assumption that Eq. (13.4) can be written in the form

$$\frac{\partial \Psi(x,t)}{\partial s} = \frac{1}{\sqrt{1+U^2}} \left[\frac{\partial}{\partial x} \left(E \frac{\partial \Psi(x,t)}{\partial x} \right) - k\Psi - k\gamma \mp S \right]$$

where

$$\frac{\partial \Psi(x,t)}{\partial s}$$

is the derivative in the direction identified by the director cosines

$$\frac{U}{\sqrt{1+U^2}} \qquad \frac{1}{\sqrt{1+U^2}}$$

If, as in Fig. 13.5, $P_1(x,t)$ and $P_2(x+l, t+\Delta t)$ are two points of the x,t plane which belong to a straight line having the defined direction, the following approximation can be adopted:

$$\Psi(x+l,t+\Delta t) - \Psi(x,t) \cong \frac{\partial \Psi(P_1)}{\partial s} \left(\sqrt{l^2 + \Delta t^2} \right) = \frac{\partial \Psi(x,l)}{\partial s} \Delta t \sqrt{l+U^2}$$

where the terms that contain higher-order infinitesimal with respect to $\Delta t\sqrt{1+U^2}$ are neglected.

Introducing the expression

$$\frac{\partial \Psi(x,l)}{\partial s} \Delta t \sqrt{1 + U^2} \cong \Psi(x + l, t + \Delta t) - \Psi(x, t)$$

Equation (13.4) is transformed into

$$\frac{\Psi(x + l, t + \Delta t) - \Psi(x, t)}{\Delta t} = \frac{\partial}{\partial x}\left(E\frac{\partial \Psi}{\partial x}\right) - k\gamma \mp S \qquad (13.29)$$

With the notation

$$\Psi^n(x) = \Psi(x, t_n)$$

the following set of initial and boundary problems can be obtained, for $0 \le x \le L$:

$$\begin{cases} \dfrac{d}{dx}\left(E\dfrac{\partial \Psi^n(x)}{\partial x}\right) - (1 + k\Delta t)\Psi^n(x) = \Phi^{n-1} \\[2mm] \Psi^n(0) = 0, \qquad \dfrac{d\Psi^n(L)}{dx} = 0 \\[2mm] \Psi^0(x) = 0 \end{cases} \qquad (13.30)$$

in the [0,L] length and denoting, for simplicity,

$$\Phi^{n-1}(x) = \Psi^{n-1}(x - l) + (-k\gamma \pm S)\Delta t$$

As pointed out in the work of Douglas and Russel already quoted, and Strang and Fix (1973), the solution of problem (13.29), for every x and n, is equivalent to the minimum of the *functional*

$$\sum \Psi(x) = \int_0^L \left\{ E\Delta t \left[\frac{d\Psi(x)}{dx}\right]^2 + (1 + k\Delta t)[\Psi(x)]^2 - 2\Phi^{n-1}(x)\Psi(x) \right\} dx$$

$$(13.31)$$

which is valid in the whole set of the identifiable continuous functions $\Psi(x)$, being null at $x = 0$, for which the square of the first derivative can be summed in the [0,L] interval. In fact, following the general theory, with considerations omitted in this description (Zienkiewicz 1972), it can be proved that, if a restricted function $\Psi n(x)$ complies with the condition

$$\sum (\Psi^n) \le \sum (\Psi)$$

for any possible value of Ψ, such a function can be also the solution of the problems that satisfy the conditions (13.30).

It is worthy to add that, for this *variational* formulation, it is not necessary that the condition

$$\frac{d\Psi^n(L)}{dx} = 0$$

should be posed a priori, being the consequence of the minimum for the solution of $\Psi^n(x)$. It is also worthy to remember that, in order to find an approximate solution for the problem of minimum (13.31) by means of linear finite elements, the $[0,L]$ length should be divided into M intervals, each one $l = U\Delta t$ long, recalling the family of functions $\phi_j(x)$ introduced in (13.8) and (13.9).

Following the above outlines of the general theory, which is better described in the textbooks of numerical methods, in the case of $M \geq 2$, the approximate solution to the problem of minimum (13.31) can be found in those particular linear combinations

$$\Psi^n(x) = \sum_{j=1}^{M+1} a_j^n \varphi_j$$

which satisfy the boundary conditions $a_0^n = 0, a_{M+1}^n = a_M^n$. The problem unknowns are, therefore, the set of $a_1^n, a_2^n, \ldots, a_M^n$, or the vector $\mathbf{Q}^n$ already defined in (13.24), for which the function

$$\Psi^n(x) = \sum_{j=1}^{M} a_j^n \varphi_j$$

minimises the integral (13.31), in which the last term is

$$\Phi^{n-1}(x) = \Psi^{n-1}(x-l) - k\gamma\Delta t \pm S\Delta t \quad \text{for} \quad n = 1, 2, \ldots M$$

and for which

$$\Psi^0(x) = 0 \qquad\qquad \text{in} \quad [0,L]$$

Integral (13.31) can be analysed considering separately the various terms. Concerning the first term, it follows that

$$\int_0^L \left(\frac{d\Psi^n(x)}{dx}\right)^2 dx = \sum_{j=1}^{M} \left(\int_{x_{j-1}}^{x_j} \left[\frac{d\Psi^n(x)}{dx}\right]^2 dx \right)$$

$$= \frac{1}{l}\left[(a_1^n)^2 + \sum_{j=2}^{M} \left(a_j^n - a_{j-1}^n\right)^2 \right]$$

(13.32)

Introducing the column vector

$$\mathbf{Q}^{n^T} = \left[a_0^n, a_1^n, \ldots a_M^n\right]^{\mathrm{T}} \tag{13.33}$$

with $a_0^n = 0$, and the matrix $\mathbf{A}$ defined in (13.21), expression (13.32) can be written in vector-matrix form

$$\int_0^L \left(\frac{\mathrm{d}\Psi^n(x)}{\mathrm{d}x}\right)^2 \mathrm{d}x = \frac{1}{l}\mathbf{Q}^{n^T}\mathbf{A}\mathbf{Q}^n$$

Concerning the second term in the right-hand side of (13.31), for $j = 1$ it becomes

$$\int_0^{x_1} (\Psi^n(x))^2 \mathrm{d}x = \left(a_1^n\right)^2 \int_0^l \frac{x^2}{l^2}\mathrm{d}x = \frac{l}{3}\left(a_1^n\right)^2$$

and for $j = 2, 3, \ldots M$

$$\int_{x_{j-1}}^{x_j} (\Psi^n(x))^2 \mathrm{d}x = \int_{x_{j-1}}^{x_j} \left[a_{j-1}^n\left(1 - \frac{x - x_{j-1}}{l}\right) + a_j^n\left(\frac{x - x_{j-1}}{l}\right)\right]^2 \mathrm{d}x$$

$$= \frac{1}{3}\left[\left(a_{j-1}^n\right)^2 + \left(a_{j-1}^n\right)\left(a_j^n\right) + \left(a_j^n\right)^2\right] = \frac{l}{6}\begin{bmatrix} 2 & 1 \\ 1 & 2 \end{bmatrix}\begin{pmatrix} a_{j-1}^n \\ a_j^n \end{pmatrix}$$

or

$$\int_0^{L_1} (\Psi^n(x))^2 \mathrm{d}x = \frac{l}{3}\left(a_1^n\right)^2 + \frac{l}{6}\sum_{j=2}^M \left(a_{j-1}^n, a_j^n\right)\begin{bmatrix} 2 & 1 \\ 1 & 2 \end{bmatrix}\begin{pmatrix} a_{j-1}^n \\ a_j^n \end{pmatrix}$$

which, taking into consideration the M-order matrix $\mathbf{B}$ defined in (13.22), becomes, in vector matrix form,

$$\int_0^L (\Psi)^2\mathrm{d}x = \frac{l}{6}\mathbf{Q}^{n^T}\mathbf{B}\mathbf{Q}^n \tag{13.34}$$

with the vectors $\mathbf{Q}^n$ and $\mathbf{Q}^{n^T}$ already defined, respectively, in (13.24) and (13.33).

Finally, for the third term of (13.31), it is convenient to take into consideration the following two integrals:

$$I_1 = \int_0^L \Psi^{n-1}(x-l)\Psi^n(x)dx \tag{13.35a}$$

and

$$I_2 = \int_0^L \Psi^n(x)dx \tag{13.35b}$$

giving the linear combination

$$\int_0^L \Phi^{n-1}(x)\Psi^n(x)dx = I_1 - I_2(\gamma k \mp S)\Delta t \tag{13.36}$$

Setting

$$\hat{\mathbf{Q}}^{(n-1)T} \equiv \left(\hat{a}_1^{n-1}, \hat{a}_2^{n-1}, \ldots, \hat{a}_{M-1}^{n-1}\right)^T = \left(a_0^{n-1}, a_1^{n-1}, \ldots a_{M-1}^{n-1}\right)$$

the integral of (13.35a) becomes

$$I_1 = \int_0^L \Psi^{n-1}(x-l)\Psi^n(x)dx = \frac{l}{6}\hat{\mathbf{Q}}^{(n-1)T}\mathbf{B}\mathbf{Q}^n \tag{13.37}$$

being $\mathbf{B}$ the matrix (13.22) and $\mathbf{Q}^n$ the vector (13.24); similarly, for the integral in (13.35b),

$$I_2 = \int_0^L \Psi^n(x)dx = \frac{l}{2}a_1^n + \frac{1}{2}\sum_{j=2}^{M-1}\left(a_j^n + a_{j+1}^n\right)$$

which, being $a_0^n = 0$, can be written in the form

$$I_2 = \frac{l}{2}\sum_{j=1}^M \left(a_j^n + a_{j-1}^n\right)$$

After combining in vector-matrix form the preceding expressions, the calculation of Ψ^n is moved to the identification of the vector $\mathbf{Q}^n$ that minimises the quadratic function

$$\sum_{n}(\mathbf{Q}^n) = \mathbf{Q}^{n\mathrm{T}}\mathbf{A}\mathbf{Q}^n - \frac{1}{3}\hat{\mathbf{Q}}^{(n-1)\mathrm{T}}\mathbf{B}\mathbf{Q}^n - 2(k\gamma - S)\Delta t\frac{1}{2}\sum_{j=1}^{M}\left(a_j^n + a_{j-1}^n\right) \quad (13.38)$$

being

$$\mathbf{A} = \Delta t E\mathbf{A} + (1 + k\Delta t)\mathbf{C}$$

and

$$\mathbf{A}\mathbf{Q}^n = 1$$

After deriving expression (13.38) and putting equal to zero the derivative, the problem is transformed into the linear system, written in vector-matrix form

$$\left[\Delta t\frac{E}{l}\mathbf{A} + \frac{l(1+k\Delta t)}{6}\mathbf{B}\right]\mathbf{Q}^n - \frac{l}{6}\mathbf{Q}^{n-1}\mathbf{B} + k\gamma\Delta t\mathbf{H} = 0 \quad (13.39a)$$

or

$$\left[\frac{E}{U}\mathbf{A} + \frac{l(1+k\Delta t)}{6}\mathbf{B}\right]\mathbf{Q}^n - \frac{l}{6}\mathbf{Q}^{n-1}\mathbf{B} + k\gamma\Delta t\mathbf{H} = 0 \quad (13.39b)$$

where

$$U\Delta t = l$$

and

$$\mathbf{H} = \left(l, l, \ldots\frac{l}{2}\right)$$

The following example can help to understand the described procedure. Considering a four space-step stream ($M + 1 = 4$), the expression (13.39a) is

$$\frac{\Delta t E}{l}\begin{vmatrix} 2 & -1 & 0 & 0 \\ -1 & 2 & -1 & 0 \\ 0 & -1 & 2 & -1 \\ 0 & 0 & -1 & 1 \end{vmatrix} + \frac{lk}{6}\begin{vmatrix} 4 & 1 & 0 & 0 \\ 1 & 4 & 1 & 0 \\ 0 & 1 & 4 & 1 \\ 0 & 0 & 1 & 2 \end{vmatrix}\begin{bmatrix} a_1^n \\ a_2^n \\ a_3^n \\ a_4^n \end{bmatrix}$$

$$= \frac{l}{6}\begin{vmatrix} 4 & 1 & 0 & 0 \\ 1 & 4 & 1 & 0 \\ 0 & 1 & 4 & 1 \\ 0 & 0 & 1 & 2 \end{vmatrix}\begin{bmatrix} a_0^{n-1} \\ a_1^{n-1} \\ a_2^{n-1} \\ a_3^{n-1} \end{bmatrix} - k\gamma\Delta t\begin{bmatrix} l \\ l \\ 0.5l \end{bmatrix}$$

After some algebraic operations, the explicit form of this system is

$$a_1^n \left(-\frac{\Delta tE}{l} + \frac{l\kappa}{6} \right) + a_2^n \left(\frac{2\Delta tE}{l} + \frac{4l\kappa}{6} \right) + a_3^{n-1} \left(-\frac{\Delta tE}{l} + \frac{l\kappa}{6} \right)$$

$$= \frac{l}{6} \left(a_0^{n-1} + 4a_1^{n-1} + a_2^{n-1} \right) - k\gamma l\Delta t$$

$$a_1^n \left(\frac{2\Delta tE}{l} + \frac{4l\kappa}{6} \right) + a_2^n \left(-\frac{\Delta tE}{l} + \frac{l\kappa}{6} \right) = \frac{l}{6} \left(4a_0^{n-1} + a_1^{n-1} \right) - k\gamma l\Delta t \qquad (13.40)$$

$$\cdots\cdots\cdots\cdots\cdots\cdots\cdots\cdots\cdots\cdots\cdots\cdots\cdots\cdots\cdots\cdots\cdots\cdots\cdots$$

$$a_M^n \left(-\frac{\Delta tE}{l} + \frac{l\kappa}{6} \right) + a_{M+1}^n \left(\frac{\Delta tE}{l} + \frac{2l\kappa}{6} \right) = \frac{l}{6} \left(a_{M-1}^{n-1} + 2a_M^{n-1} \right) - \frac{k\gamma l\Delta t}{2}$$

where, for simplicity, is $\kappa = (1 + \gamma\Delta t)$.

For any time interval n, the solution of the system (13.40) is

$$a_1^n, a_2^n, \ldots a_{M+1}^n$$

as function of the values

$$a_0^{n-1}, a_1^{n-1}, \ldots a_M^{n-1}$$

obtained at the preceding time interval $n - 1$.

The fundamental terms of the problem must be chosen in order to satisfy the Péclet and CFL conditions

$$\frac{\Delta xU}{E} \leq 2 \qquad \frac{\Delta tU}{\Delta x} \leq 1$$

which are essential to obtain a stable and significant solution.

13.4 An Application

The procedure described in the preceding lines is applied to a river stretch subject to a continuous pollutant injection of infinite duration at the cross section $x = 0$. It is a case already examined in preceding chapters. There is no local pollutant injection.

The characteristic data of the problem are

$\Delta x = l =$	10 m
$U =$	0.5 m/s
$\Delta t =$	20 s
$E =$	3.0 m²/s
$C(0,0) = \gamma$	100%

Table 13.1 Original values
at $t = 0$

$a(0,0)$	0.00	$a(4,0)$	-100.00
$a(1,0)$	-100.00	$a(5,0)$	-100.00
$a(2,0)$	-100.00	$a(6,0)$	-100.00
$a(3,0)$	-100.00	$a(7,0)$	-100.00

Table 13.2 Solution at time
step $t = 1$

$a(1,1)$	-26.59	$a(5,1)$	-99.61
$a(2,1)$	-72.84	$a(6,1)$	-99.91
$a(3,1)$	-93.42	$a(7,1)$	-99.98
$a(4,1)$	-98.41	$a(8,1)$	-99.99

The case is first examined for nonconservative pollutant, with $k = 0.001 \text{ s}^{-1}$.

The stream is divided into eight elements of 10 m long and the pollutant injection is examined for 12 time intervals. The data are chosen so that they satisfy a priori the CFL condition and, secondarily, the Péclet condition, being

$$\frac{\Delta t U}{\Delta x} = 1 \leq 1$$

and

$$\frac{U \Delta x}{E} = 1.67 \leq 2$$

The original values, at time instant $t = 0$, are in Table 13.1, at the considered space intervals:where we have adopted the notation

$$a(x, nt) = a_x^{nt}$$

With the known values for $t = 0$ and the other original values for the $t = 1$, the system (13.40) becomes

$$
\begin{array}{llll}
18.8\ a(1,1) & -4.3\ a(2,1) & = & -187.00 \\
-4.3\ a(1,1) & +18.8\ a(2,1) & -4.3\ a(3,1) & = & -853.00 \\
-4.3\ a(2,1) & +18.8\ a(3,1) & -4.3\ a(4,1) & = & -1020.00 \\
-4.3\ a(3,1) & +18.8\ a(4,1) & -4.3\ a(5,1) & = & -1020.00 \\
-4.3\ a(4,1) & +18.8\ a(5,1) & -4.3\ a(6,1) & = & -1020.00 \\
-4.3\ a(5,1) & +18.8\ a(6,1) & -4.3\ a(7,1) & = & -1020.00 \\
-4.3\ a(6,1) & +18.8\ a(7,7) & -4.3\ a(8,1) & = & -1020.00 \\
-4.3\ a(7,1) & +9.4\ a(8,1) & & = & -510.00
\end{array}
$$

The solution is presented in Table 13.2, obtained with the application of the conventional solution of linear systems.

Table 13.3 Concentration values at time step $t = 1$

$C(1,1)$	73.41	$C(5,1)$	0.39
$C(2,1)$	27.16	$C(6,1)$	0.09
$C(3,1)$	6.58	$C(7,1)$	0.02
$C(4,1)$	1.59	$C(8,1)$	0.01

Table 13.4 Solution at time step $t = 2$

$a(1,2)$	−11.3	$a(5,2)$	−95.3
$a(2,2)$	−34.4	$a(6,2)$	−98.5
$a(3,2)$	−65.1	$a(7,2)$	−99.5
$a(4,2)$	−86.1	$a(8,2)$	−99.8

Table 13.5 Concentration at time step $t = 2$

$C(1,2)$	88.7	$C(5,2)$	4.7
$C(2,2)$	65.6	$C(6,2)$	1.5
$C(3,2)$	34.9	$C(7,2)$	0.5
$C(4,2)$	13.9	$C(8,2)$	0.2

By recalling (13.10), the relative pollutant concentration is in Table 13.3

The values of $a(1,1)$, $a(2,1)$,....$a(7,1)$, with $a(0,1) = 0$ are used to construct the right-hand side of the system for the time step $t = 2$, while the coefficients of the left-hand side remain unchanged, giving (values are written with only one decimal point)

$$
\begin{array}{rrrcr}
18.8\ a(1,2) & -4.3\ a(2,2) & & = & 64.3 \\
-4.3\ a(1,2) & +18.8\ a(2,2) & -4.3\ a(3,2) & = & 318.7 \\
-4.3\ a(2,2) & +18.8\ a(3,2) & -4.3\ a(4,2) & = & 705.6 \\
-4.3\ a(3,2) & +18.8\ a(4,2) & -4.3\ a(5,2) & = & 928.2 \\
-4.3\ a(4,2) & +18.8\ a(5,2) & -4.3\ a(6,2) & = & 997.8 \\
-4.3\ a(5,2) & +18.8\ a(6,2) & -4.3\ a(7,2) & = & 1014.6 \\
-4.3\ a(6,2) & +18.8\ a(7,2) & -4.3\ a(8,2) & = & 1018.7 \\
-4.3\ a(7,2) & +9.4\ a(8,2) & & = & 509.7 \\
\end{array}
$$

The solution for the unknowns $a(x,2)$ is presented in Table 13.4 and the concentration in Table 13.5.

The solution for the other time steps can be found proceeding in the same way. For the 12 time steps considered, the result is given in Table 13.6. The pollutant reaches a stable condition after a time interval that depends on the distance from the injection, as already shown in preceding sections.

Also in this case, it is of interest to compare this solution with that of an analytical procedure, as shown in Fig. 13.6.

The same problem is examined for a conservative pollutant, with the same original data of the preceding case but assuming $k = 0.00\ \text{s}^{-1}$.

Table 13.6 Concentration for the injection of a nonconservative pollutant

$i \backslash j$	0	1	2	3	4	5	6	7	8	9	10	11
0	100.0	100.0	100.00	100.0	100.00	100.00	100.0	100.00	100.00	100.0	100.00	100.00
1	0.0	73.41	88.71	93.74	95.81	96.77	97.25	97.50	97.64	97.71	97.76	97.78
2	0.0	27.16	65.58	81.65	88.75	92.15	93.87	94.79	95.29	95.57	95.73	95.82
3	0.0	6.58	34.91	61.55	76.49	84.41	88.67	91.00	92.31	93.06	93.49	93.74
4	0.0	1.59	13.94	37.85	58.87	72.53	80.70	85.49	88.29	89.92	90.89	91.45
5	0.0	0.39	4.71	18.97	39.16	56.89	69.47	77.62	82.71	85.83	87.71	88.84
6	0.0	0.09	1.47	8.12	22.46	40.12	55.79	67.48	75.45	80.60	83.84	85.83
7	0.0	0.02	0.46	3.28	11.82	26.28	42.57	56.81	67.51	74.88	79.69	82.72
8	0.0	0.01	0.24	1.92	8.01	20.20	35.91	50.98	62.99	71.57	77.30	80.96

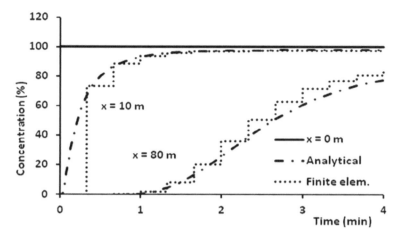

Fig. 13.6 Solution for the injection of a nonconservative pollutant, at two cross sections of the river stretch

With these data, for the time step $t = 1$, the system (13.40) becomes

$$
\begin{array}{llll}
18.7\ a(1,1) & -4.3\ a(2,1) & & = & -166.7 \\
-4.3\ a(1,1) & +18.7\ a(2,1) & -4.3\ a(3,1) & = & -833.3 \\
-4.3\ a(2,1) & +18.7\ a(3,1) & -4.3\ a(4,1) & = & -1000.0 \\
-4.3\ a(3,1) & +18.7\ a(4,1) & -4.3\ a(5,1) & = & -1000.0 \\
-4.3\ a(4,1) & +18.7\ a(5,1) & -4.3\ a(6,1) & = & -1000.0 \\
-4.3\ a(5,1) & +18.7\ a(6,1) & -4.3\ a(7,1) & = & -1000.0 \\
-4.3\ a(6,1) & +18.7\ a(7,1) & -4.3\ a(8,1) & = & -1000.0 \\
-4.3\ a(7,1) & +9.3\ a(8,1) & & = & -500.0
\end{array}
$$

Proceeding in the same way as before, the final solution, for 11 time steps and eight elements on the x-axis, is shown in Table 13.7.

The pollutant behaviour is shown in Fig. 13.7, for two river cross sections ($x = 10$ m and $x = 80$ m), and the finite element solution is compared with the analytical solution. It is worthy to notice that the FEM gives values that match up to those given by the analytical method, but as far as the distance from the injection increases, there is a clear trend for the final element to anticipate the value of the resulting pollutant concentration.

The observed discrepancies can be explained by the general statement that the continuum cannot be entirely interpreted by means of a discontinuous approach.

The exhaustive answer to this interpretation should be that of the experimental measurement in the real field, at the calibration and validation step of the model.

13.5   The Pollutant Wave

The approach described in the preceding paragraphs is useful to deal with more complex cases, in a way closer to the most frequent real conditions.

Table 13.7 Concentration for the injection of a conservative pollutant

Label	Value	Label	Value	Label	Value	Label	Value	Label	Value	Label	Value
C(0,0)	100.00	C(0,2)	100.00	C(0,4)	100.00	C(0,6)	100.00	C(0,8)	100.00	C(0,10)	100.00
C(1,0)	0.00	C(1,2)	90.10	C(1,4)	97.69	C(1,6)	99.31	C(1,8)	99.77	C(1,10)	99.92
C(2,0)	0.00	C(2,2)	67.26	C(2,4)	91.78	C(2,6)	97.49	C(2,8)	99.15	C(2,10)	99.70
C(3,0)	0.00	C(3,2)	36.17	C(3,4)	80.04	C(3,6)	93.51	C(3,8)	97.75	C(3,10)	99.19
C(4,0)	0.00	C(4,2)	14.63	C(4,4)	62.23	C(4,6)	86.25	C(4,8)	95.00	C(4,10)	98.17
C(5,0)	0.00	C(5,2)	5.01	C(5,4)	41.80	C(5,6)	75.09	C(5,8)	90.30	C(5,10)	96.33
C(6,0)	0.00	C(6,2)	1.59	C(6,4)	24.24	C(6,6)	60.94	C(6,8)	83.43	C(6,10)	93.48
C(7,0)	0.00	C(7,2)	0.51	C(7,4)	12.93	C(7,6)	46.98	C(7,8)	75.48	C(7,10)	89.98
C(8,0)	0.00	C(8,2)	0.26	C(8,4)	8.85	C(8,6)	39.89	C(8,8)	70.83	C(8,10)	87.83
C(0,1)	100.00	C(0,3)	100.00	C(0,5)	100.00	C(0,7)	100.00	C(0,9)	100.00	C(0,11)	100.00
C(1,1)	74.30	C(1,3)	95.43	C(1,5)	98.76	C(1,7)	99.60	C(1,9)	99.86	C(1,11)	99.95
C(2,1)	27.76	C(2,3)	84.14	C(2,5)	95.53	C(2,7)	98.56	C(2,9)	99.50	C(2,11)	99.82
C(3,1)	6.84	C(3,3)	64.06	C(3,5)	88.73	C(3,7)	96.21	C(3,9)	98.66	C(3,11)	99.51
C(4,1)	1.68	C(4,3)	39.78	C(4,5)	77.12	C(4,7)	91.72	C(4,9)	96.98	C(4,11)	98.89
C(5,1)	0.41	C(5,3)	20.17	C(5,5)	61.11	C(5,7)	84.37	C(5,9)	94.02	C(5,11)	97.75
C(6,1)	0.10	C(6,3)	8.75	C(6,5)	43.53	C(6,7)	74.20	C(6,9)	89.55	C(6,11)	95.97
C(7,1)	0.03	C(7,3)	3.59	C(7,5)	28.84	C(7,7)	63.11	C(7,9)	84.18	C(7,11)	93.74
C(8,1)	0.01	C(8,3)	2.13	C(8,5)	22.33	C(8,7)	56.96	C(8,9)	80.94	C(8,11)	92.35

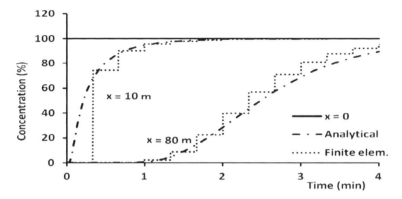

Fig. 13.7 Solution for the injection of a conservative pollutant, at two cross sections of the river

Fig. 13.8 Concentration at the initial cross section of the stream ($x = 0$), as a function of time

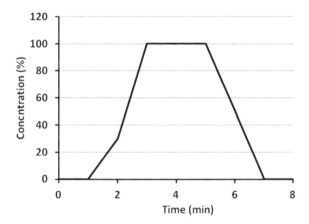

In the following lines, an application is presented with a pollutant injection in the form of a "wave", like that Fig. 13.8, at the initial cross section of the stream.

A one-dimensional stream is considered with the following data:

$$U = 0.15\text{m/s} \qquad k = 0.00 \text{ s}^{-1}$$
$$E = 5\text{m}^2/\text{s} \qquad \Delta t = 60 \text{ s}$$
$$\Delta x = l = U\Delta t = 9 \text{ m}$$

The total time interval $[0, T]$ is divided into N elementary steps and the stream into $M + 1$ elements. For this particular case, the original system (13.40) is modified in the form

$$a_1^n\left(\frac{2\Delta tE}{1}+\frac{4l\kappa}{6}\right)+a_2^n\left(-\frac{\Delta tE}{1}+\frac{l\kappa}{6}\right)=\frac{1}{6}\left(4a_0^{n-1}+a_1^{n-1}\right)-k\gamma l\Delta t$$

$$a_1^n\left(-\frac{\Delta tE}{l}+\frac{l\kappa}{6}\right)+a_2^n\left(\frac{2\Delta tE}{l}+\frac{4l\kappa}{6}\right)+a_3^{n-1}\left(-\frac{\Delta tE}{l}+\frac{l\kappa}{6}\right)$$

$$=\frac{l}{6}\left(a_0^{n-1}+4a_1^{n-1}+a_2^{n-1}\right)-k\gamma l\Delta t$$

....

$$a_M^n\left(-\frac{\Delta tE}{l}+\frac{l\kappa}{6}\right)+a_{M+1}^n\left(\frac{\Delta tE}{l}+\frac{l\kappa}{3}\right)=\frac{l}{6}\left(a_{M-1}^{n-1}+2a_M^{n-1}\right)-\frac{k\gamma l\Delta t}{2}$$

where, unlike the case examined in the preceding paragraphs, the known values a_j^n to be inserted in the right-hand side are defined by the specific value of γ_n characteristic of the time element considered at any time step.

In the initial 0–1 time element, being the concentration $C(0,0) = 0.00$, according to the definition (13.3), it follows that $\gamma_1 = 0$ and, being zero the concentration at all the other elements, the solution of the system is

$$a_0^0 = 0.00$$
$$a_1^0 = 0.00$$
$$a_2^0 = 0.00$$

Inserting these known values in the right-hand side, the solution of the system gives the values at the end of the interval ($t = 1$), namely,

$$a_1^1 = 0.00$$
$$a_2^1 = 0.00$$

from which, according to (13.3), the concentrations

$$C(1, 1) = a_1^1 + \gamma_1 = 0.00$$
$$C(2, 1) = a_2^1 + \gamma_1 = 0.00$$

In the time element 1–2, the initial value is $C(0,1) = 0.00$; therefore, $\gamma_2 = 0.00$. The known values, to insert in the right-hand side of the system, are

$$a_0^1 = C(0, 1) - \gamma_2 = 0.00$$
$$a_1^1 = C(1, 1) - \gamma_2 = 0.00$$
$$a_2^1 = C(2, 1) - \gamma_2 = 0.00$$

and with these values the solution is

$$a_1^2 = 0.00$$
$$a_2^2 = 0.00$$
$$a_3^2 = 0.00$$

then the concentration at time instant $t = 2$ in the various stream elements

$$C(1,2) = a_1^2 + \gamma_2 = 0.00$$
$$C(2,2) = a_2^2 + \gamma_2 = 0.00$$
$$C(3,2) = a_3^2 + \gamma_2 = 0.00$$

At the beginning of the time interval [2–3], the concentration at the initial cross section, $x = 0$, becomes $C(0,2) = 30.00$, then $\gamma_3 = 30.00$. The known values inserted in the right-hand side become

$$a_0^2 = C(0,2) - \gamma_3 = 0.00$$
$$a_1^2 = C(1,2) - \gamma_3 = -30.00$$
$$a_2^2 = C(2,2) - \gamma_3 = -30.0$$

and the solution

$$a_1^3 = -7.58$$
$$a_2^3 = -15.89$$
$$a_3^3 = -21.63$$

with the concentration

$$C(1,3) = a_1^3 + \gamma_3 = 22.42$$
$$C(2,3) = a_2^3 + \gamma_3 = 14.11$$
$$C(3,3) = a_3^3 + \gamma_3 = 8.37$$

In time interval [3,4], the concentration at the initial cross section $x = 0$ is $C(0,3) = 100.00$, then $\gamma_4 = 100.00$; therefore, the known values inserted in the right-hand side of the system are

$$a_0^3 = C(0,3) - \gamma_4 = 0.00$$
$$a_1^3 = C(1,3) - \gamma_4 = 77.58$$
$$a_2^3 = C(2,3) - \gamma_4 = 85.89$$

The solution is

$$a_1^4 = 21.60$$
$$a_2^4 = 45.65$$
$$a_3^4 = 63.93$$

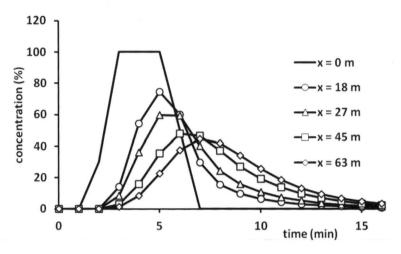

Fig. 13.9 Pollutant concentration, as a function of time, for eight elements of a stream

And, therefore, the concentrations are

$$C(1,4) = a_1{}^4 + \gamma_4 = 78.40$$
$$C(2,4) = a_2{}^4 + \gamma_4 = 54.35$$
$$C(3,4) = a_3{}^4 + \gamma_4 = 36.07$$

The procedure can be repeated for the next time intervals, assuming a value of $\gamma_n = C(0,t_n)$ for any known concentration at the initial cross section of the stream.

Figure 13.9 shows the result for eight consecutive elements of the stream. The deformation of the polluting "wave" is evident, as it is also the progressive abatement of the peak. The example confirms that the numerical method can be useful to simulate complex situations of pollutant injection and behaviour in the stream. As in the preceding sections, the application is carried out by means of a common spreadsheet, in order to explain how the various terms of the problems have to be manipulated for the computation. Obviously, the described formulations can be developed using more refined computer packages, with an increased capability of analysing the most frequently encountered practical conditions.

13.6 Additional Comments on the Finite Element Method

The finite element method is now one of the most advanced methods to solve differential equations in discrete form. The preceding pages, with the examples, show how it can be adapted to the some common cases of river pollution and how it can produce a realistic response. Some complexities still characterise the method, especially in its initial steps, when its mathematical fundamentals have to be understood and tailored to the reality of the water body, but as soon as the algorithm has been translated into a form suitable to be filled with the relevant data, the finite element method is able to give a prompt solution. Like the finite difference method, it enables the fundamental differential equations to be transformed into a system of ordinary linear equations.

One of the main causes of concern is the simultaneous presence of parabolic and hyperbolic terms in the fundamental differential equation which are discretised. As pointed out, they require different mathematical procedures, which are not easy to handle inside the same formal expression. Moreover, it has to comply with some restrictions involving the basic data, which, aggregated in some expressions, should satisfy strict constraints. Therefore, the used data should remain inside some pre-established intervals, which, very often, have no real meaning in relation with the scale of simulation adopted for the model. This occurs, in particular, for the length Δx and time Δt intervals, which must keep reasonable values: if too short, they should consider details not always available and result in the increase of computational time; if too long, they could exclude particular important details of the river and decrease of accuracy, which may be significant in the various processes of pollutant transport.

The Ritz-Galerkin approach is an example of how to separate the advection from the dispersion terms (Heinrich et al. 1977; Seo et al. 2005). In effect, it can give a stable and acceptable solution if Δt and Δx are *sufficiently small*. In particular, the smaller Δx is, the more acceptable is the solution. Such a condition is not always consistent with the river reality, in which a very short reach is difficult to identify and can lose significance in comparison with the size of the other parameters that describe the water body.

Reducing the integration instability is still the primary goal of advanced mathematics and several solutions are proposed, with application to real cases. They still belong to high scientific research and require refinements before becoming a tool of current practical procedure.

If compared with the finite difference method, the finite element method can appear more difficult to handle. However, the positive aspects of the latter can be appreciated in the most complex cases, when the application of the FDM could lead to numerous repeated equations, difficult to handle even with the most advanced computing equipment.

In the scientific field, there are still items for its improvement and its more effective application in the future. Nevertheless, at the present time, the FEM seems to be well developed and sufficiently consolidated to be applied in practice, favoured by computing procedures and packages that make its application easy. Useful information concerning the method and its applications can be found in several books of this series (e.g. Vreugdenhil 1994; Scarlatos 1996; Szymkiewicz 2010).

References

Al-Lavatia M, Sharply RC, Wang H (1999) Second order characteristic methods for advection–diffusion equations and comparison to schemes. Adv Water Res 22:741–768

Casulli V, Cheng RT (1992) Semi-implicit finite differences methods for three dimensional shallow water flow. Int J Numer Method Fluid 15:629–648

Celia MA, Russel TF, Herrera I, Ewing RE (1990) An Eulerian–Lagrangian localised adjoint method for the advection–diffusion equation. Adv Water Res 13(4):187–206

Douglas J, Russel TF (1982) Numerical methods for convection-dominated diffusion problems based on combining the method of characteristics with finite element or finite difference procedures. SIAM J Numer Anal 19:871–885

Evans LC (2000) Partial differential equations. GSM of the American Mathematical Society, 19, Providence, RI, USA

Gelinas RJ, Doss SK (1981) The moving finite element method: application to general partial differential equations with multiple large gradient. J Comput Phys 40:202–249

Heinrich JC, Huyakorn PS, Zienkjewicz OC (1977) An 'Upwind' finite element scheme for two-dimensional convective transport equation. Int J Numer Method Eng 11:131–143

Lee ME, Seo W (2007) 2D finite element modeling of pollutant transport in Tidal River. In: Proceedings of the 32nd IAHR congress, Venezia, Italy, vol 2, p 498 (printed summary).

Li YS, Chen CP (1989) An efficient split-operator scheme for 2-D advection–diffusion simulations using finite elements and characteristics. Appl Math Model 13:248–253

O'Neil K (1981) Highly efficient, oscillation free solution of the transport equation over long times and large spaces. Water Resour Res 17(6):1665–1675

Przemienieki JS (1981) Theory of matrix structural analysis. Wiley, New York

Scarlatos P (1996) Estuarine hydraulics. In: Singh VP, Hager WH (eds) Environmental hydraulics. Water sciences and technology library. Kluwer Academic Publisher, Dordrecht, pp 289–348

Seo IW, Lee ME, Kim YH (2005) Finite element mass transport model using dispersion tensor as coefficients. In: Proceedings of the 31st IAHR congress, Seoul, Korea, vol 1, p 534 (printed summary)

Strang G, Fix GJ (1973) An analysis of finite element method. Prentice Hall, Upper Saddle River

Szymkiewicz R (2010) Numerical modeling in open channel hydraulics, vol 83, Water sciences and technology library. Springer, Dordrecht

Vreugdenhil CB (1994) Numerical methods for shallow-water flow, Water sciences and technology library. Kluwer Academic Publishers, Dordrecht

Zienkiewicz OC (1972) Introductory lectures on the finite element method. Springer, Vienna

Chapter 14
The Finite Volume Method

Abstract Beside the finite difference and finite element methods, a new numerical method has been recently proposed, which looks to be very promising for stable and reliable solutions of the fundamental differential equation of pollutant transport. It is the finite volume method, which is the object of advanced research, in view of the improvements that are necessary in order to make the method suitable for more extensive applications.

14.1 Basic Concepts

Beside the finite difference and finite element, a new promising method, the *finite volume method* (FVM), is now proposed for the numerical integration of the fundamental differential equation of pollutant transport. The method is currently applied to several practical cases, characterised by a *conservation law*. Many natural phenomena, including the pollution transport, can be expressed in conservative form, meaning that an entity varying inside a region is assessed by its amount that crosses the boundary.

The method can only be presented briefly in this chapter. The estimated fundamentals of the method are summarised in the following paragraphs. More details can be found in the abundant literature pertaining to this subject.

In a field in which the entity is identified by a proper coordinate system, the partial differential equation is discretised into cells, namely, the finite volumes. Size and shape of each cell can be chosen arbitrarily, remembering that the smaller is the cell, the higher is the accuracy. Inside the cell, the entity can be evaluated at a point with its average value over the entire cell volume. The cell boundary is a face, through which a variation of the entity can be identified, with the assumption that the value lost by a cell is gained by the contiguous one. Proper analytical transformations provide to structure the algorithm in a way that, ultimately, a linear system of equation can be developed, which is solved by means of the well-known methods

M. Benedini and G. Tsakiris, *Water Quality Modelling for Rivers and Streams*,
Water Science and Technology Library 70, DOI 10.1007/978-94-007-5509-3_14,
© Springer Science+Business Media Dordrecht 2013

Fig. 14.1 Basic layout for
the application of the FVM to
one-dimensional pollutant
transport

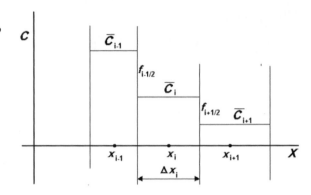

Fig. 14.1 Basic layout for the application of the FVM to one-dimensional pollutant transport

of linear algebra. The cells can have one or more dimensions, and then the method is applicable to problems of different nature.

The above concepts can be better understood by means of the following example.

Considering a one-dimensional simplified case of conservative pollutant transport with no concentrated terms, the fundamental equation is written as

$$\frac{\partial C}{\partial t} + u\frac{\partial C}{\partial x} - E\frac{\partial^2 C}{\partial x^2} = 0 \tag{14.1}$$

Putting

$$uC - E\frac{\partial C}{\partial x} = f$$

Equation (14.1) becomes

$$\frac{\partial C}{\partial t} + \frac{\partial f}{\partial x} = 0 \tag{14.2}$$

In Fig. 14.1, the behaviour of the pollutant concentration C is shown as a function of the length of a river stretch, x, in which some cells can be formed. In the one-dimensional case, a cell is simply a linear segment around a centre at $x = x_i$, in which the concentration at instant $t = t_1$, considered as the average value over the segment, is

$$\bar{C}_i(x, t_1) = \frac{1}{\Delta x_i} \int_{x-1/2}^{x+1/2} C(x, t_1)\mathrm{d}x \tag{14.3a}$$

In a similar way, the concentration at an instant $t = t_2$ is

$$\bar{C}_i(x, t_2) = \frac{1}{\Delta x_i} \int_{x-1/2}^{x+1/2} C(x, t_2) dx \tag{14.3b}$$

In the adopted interpretation, the boundaries of the i-th cell are identified by the edges of the segment, at location $x_i - \Delta x_i/2$ and $x_i + \Delta x_i/2$, respectively.

In Eq. (14.2), which expresses the *conservation law*, the term f is the *flux* of C; it corresponds to the segments $f_{i-1/2}$ and $f_{i+1/2}$, in Fig. 14.1 in which

$$x_{i-1/2} = x_i - \Delta x_i/2 \quad x_{i+1/2} = x_i + \Delta x_i/2$$

and

$$f_{i-1/2} = f\left[C\left(x_{i-1/2},\ t_1\right)\right] \quad f_{i+1/2} = f\left[C\left(x_{i+1/2},\ t_1\right)\right]$$

Integration of (14.2) in respect to time gives

$$C(x, t_2) = C(x, t_1) - \int_{t_1}^{t_2} \frac{\partial f}{\partial x} dt$$

and by recalling (14.3a) and (14.3b), the average value of $C(x, t_2)$ is

$$\overline{C}_i(x, t_2) = \overline{C}_i(x, t_1) - \frac{1}{\Delta x_i} \int_{x-1/2}^{x+1/2} \left\{ \int_{t_1}^{t_2} \frac{\partial f}{\partial x} dt \right\} dx \tag{14.4}$$

The general theory of the method invokes now the *divergence theorem*, to adapt to the examined case. The divergence of the flux f for the one-dimensional case can be written as

$$\nabla f = \frac{\partial f}{\partial x} \tag{14.5}$$

and its integration in the cell volume is expressed by the linear integral on the segment $[x_{i-1/2}, x_{i-1/2}]$. Equation (14.4) becomes, therefore,

$$\overline{C}_i(x, t_2) = \overline{C}_i(x, t_1) - \frac{1}{\Delta x_i} \left(\int_{t_1}^{t_2} f\left[C\left(x_{i+1/2}, t\right)\right] dt - \int_{t_1}^{t_2} f\left[C\left(x_{i-1/2}, t\right)\right] dt \right) \tag{14.6}$$

It is easy to recognise that the parenthesis in the right-hand side of (14.6) contains the integration of the flux f, acting normally on the cell boundary.

After these analytical manipulations, the two-variable fundamental Eq. (14.1) is transformed into a one-variable equation with the only variable t. The problem is now to search for a suitable way for solving the two integrals of (14.6).

A first step consists of differentiating such an equation with respect to time, using the notation already introduced, obtaining the equation

$$\frac{d\overline{C}_i}{dt} = -\frac{1}{\Delta x_i}\left[f_{1+1/2} - f_{1-1/2}\right]$$ (14.7)

and the problem is then brought to find appropriate values of the flux, using the available data and the initial conditions of the problem.

14.2   Additional Comments

Several complications are associated with the method, which may result in unstable solutions, due (in particular) to the discontinuity of the discretised terms, on the element boundary. Several computational schemes are proposed, in order to reduce the oscillations caused by such discontinuities. The research in the advanced numerical calculus has pointed out proper *essentially non-oscillatory* (ENO) and *weighted essentially non-oscillatory* (WENO) schemes, to be adopted in the method application (Shu 1998; Crnjaric-Zic et al. 2004; LeVeque 1990; Xu and Shu 2005). By means of such schemes, the flux is replaced by a polynomial, with proper weight in algebraic form, which is adapted to the known values of the problem and capable of reaching a rational solution.

With the most recent improvements (LeVeque 2002; Versteeg and Malalasekera 1995), the FVM has been proved successful in solving various problems of river hydraulics, particularly for the simulation of flow in steady and nonsteady conditions. Some applications could be mentioned for the problems of pollution in a river, at first for the case of pure advective transport (Chertock et al. 2006) and then for more complex conditions. Lin et al. (2007) have applied the method in a combination of the river flux pattern with the pollution transport, to simulate the water quality in case of an unsteady state caused by a dam failure.

The applications confirm the advantage of the FVM in giving a solution that can be valid for many practical cases. In respect to the other numerical methods, particularly the FDM, the FVM does not depend on the structure of discretisation cells and can be better adapted to the complex geometrical pattern and hydraulic condition in the river.

For more practical applications in water quality problems, further adaptation of the method may be still necessary. The method is, therefore, expected to become very useful in the near future, for solving problems of pollution transport in rivers and streams of complex configurations.

References

Chertock A, Kurganov A, Petrova G (2006) Finite-volume-particle method for models of transport of pollutant in shallow water. J Sci Comput 27(1–3):189–199

Crnjaric N, Vukovic S, Sopta L (2004) Balanced finite volume WENO and central WENO schemes for the shallow water open-channel equations. J Comput Phys 200:512–548

Leveque RJ (1990) Numerical methods for conservation laws. Birkhauser Verlag, Basel

Leveque RJ (2002) Finite volume method for hyperbolic problems. Cambridge University Press, Cambridge

Lin J, Xie Z, Yu D, Zhou J (2007) Finite-volume WENO scheme for the shallow water with transport of pollutant. In: Proceedings of the 32nd IAHR congress, Venezia, Italy, vol 1, p 187 (printed summary)

Shu CW (1998) Essentially non-oscillatory and weighted essentially non-oscillatory schemes for hyperbolic conservation laws. In: Cockburn B et al (eds) Advanced numerical approximation of nonlinear hyperbolic equations, Lecture notes in mathematics, vol 160. Springer, Berlin/New York: 325

Versteeg HK, Malalasekera W (1995) An introduction to computational fluid dynamics: the finite volume method. Addison-Wesley, Reading

Xu ZF, Shu CW (2005) Anti-diffusive flux corrections for high order finite difference WENO schemes. J Comput Phys 205(2):458–485

Chapter 15
Multidimensional Approach

Abstract Pollutant transport in water is indeed a multidimensional phenomenon, and the one-dimensional analysis of the previous chapters is only a simplified approach. While a three-dimensional analysis is more appropriate for lakes and reservoirs, the two-dimensional approach can be more efficient for a realistic simulation of the pollutant behaviour in rivers and streams, after its injection in the water body. The main features of the two-dimensional approach are described in this chapter.

15.1 The Two-Dimensional Case

The preceding chapters deal with the one-dimensional analysis of pollutant transport, suitable for rivers and streams, where the length along the axis of the stream is predominant. As already pointed out, the hydraulic characteristics can be assumed as average values on the cross section, in which also the pollutant can be considered uniformly distributed. These assumptions are the basis of the majority of scientific research efforts developed using numerical simulation (Ayyoubzadeh et al. 2004). The one-dimensional approach is fundamental also for the analytical method, which is still considered the most reliable way for solving the fundamental differential equation in simple geometrical configurations. It is indeed a common practice to compare the results produced by numerical methods with that of the analytical solution.

The assumptions made for the one-dimensional analysis are not always valid, particularly in large rivers, in which an injected pollutant requires some time before reaching an acceptable uniform distribution in the cross section. The pollutant presence is conceived in form of a two- or three-dimensional *plume* that expands on the cross section along a very long stretch of the river (Blumberg et al. 1996; Chao et al. 2009; Goltz and Roberts 1986; Casulli and Stelling 1998, 2011; Dortch et al. 1992).

In case the pollutant is injected at the centre of the stream, as in the Fig. 15.1, and behaves in a symmetric mode with respect to the central axis of the channel, the

M. Benedini and G. Tsakiris, *Water Quality Modelling for Rivers and Streams*,
Water Science and Technology Library 70, DOI 10.1007/978-94-007-5509-3_15,
© Springer Science+Business Media Dordrecht 2013

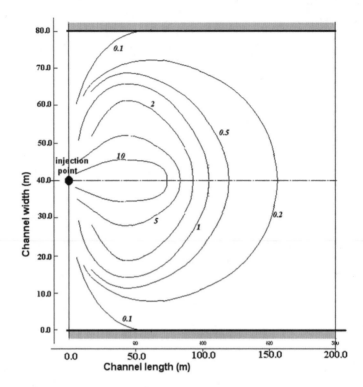

Fig. 15.1 Isoconcentration lines in the river, after a pollutant injection, at the centre of a cross section of the river

plume can be shown in form of *isoconcentration* (or equal concentration) lines. Figure 15.1 shows a typical two-dimensional case.

Moreover, the large rivers can have zones of slow stream or stagnant water, in which the pollutant transport occurs in a way different from that in the main stream. Typical is also the case of estuaries, in which the "loose boundaries" do not allow the simplified hypotheses of the one-dimensional analysis to be fulfilled.

According to these considerations, there are sufficient reasons for at least a two-dimensional analysis (Kalinowska and Rowinski 2009; Lin and Falconer 2005; Murillo et al. 2005, 2007). Normally this approach refers to a horizontal plane, where the x coordinate is taken longitudinally along the main flow direction and the y coordinate is lateral towards the embankments. At any point $P(x,y)$, the significant terms of the problem are assumed as the average values over the vertical. The case where the x,y plane is vertical and the values are the average over the transverse is less frequent in problems concerning rivers and streams.

In the next lines, it will be briefly shown how the fundamental differential equation can be transformed into discrete terms according to the two-dimensional approach. In order to achieve a solution, a complex integration process is necessary and only some simplified hints will be recalled, leaving the more complete treatment of the subject to the specialised scientific literature (Wu et al. 2001).

In a two-dimensional approach, the fundamental differential equation of pollution transport (introduced in Chap. 4) becomes

$$\frac{\partial C}{\partial t} + u\frac{\partial C}{\partial x} + v\frac{\partial C}{\partial y} = E\left(\frac{\partial^2 C}{\partial x^2} + \frac{\partial^2 C}{\partial y^2}\right) \tag{15.1}$$

in which u and v are the water velocity components along, the x- and y-axes, respectively.

An analytical solution of this equation involves several mathematical manipulations, with the final result of very complex formulations. It is, therefore, more convenient to search for numerical methods for the solution.

Transforming Eq. (15.1) into finite terms, in view of an application of the finite difference or finite element method, can be easy in principle. However, there are several conditions which should be met in order to avoid unacceptable solutions, as repeatedly mentioned in the preceding chapters. The main source of difficulty lies always on the simultaneous presence of parabolic and hyperbolic terms, which have already caused some difficulties even in the one-dimensional analysis.

There are several attempts to develop procedures for the application of numerical methods in two- or three-dimensional cases, leading to complex formulations that require severe constraints in the choice of the proper geometrical and hydraulic terms. In the following paragraphs, a short description is given to a promising procedure for a two-dimensional approach. The procedure is based on the finite difference method already mentioned in Chap. 12 (Cheng et al. 1984; Casulli and Cheng 1992). It can be formulated satisfactorily if the pollution transport occurs mainly in the direction of the flow (*transportive property*), which is indeed the most common case in pollutant transport in rivers and streams.

From the general theory dealing with two-dimensional problems, some simplified interpretations of the fundamental differential equation are now proposed. Specifically the analysis described belongs to the explicit upwind numerical scheme.

For the time derivative

$$\frac{\partial C}{\partial t} = \frac{C_{i,j}^{t+1} - C_{i,j}^{t}}{\Delta t} \tag{15.2}$$

for advection

$$u\frac{\partial C}{\partial x} + v\frac{\partial C}{\partial y} = u\frac{C_{i,j}^{t} - C_{i-1,j}^{t}}{\Delta x} + v\frac{C_{i,j}^{t} - C_{i,j-1}^{t}}{\Delta y} \tag{15.3}$$

and for dispersion

$$E\left(\frac{\partial^2 C}{\partial x^2} + \frac{\partial^2 C}{\partial y^2}\right) = E\frac{C_{i-1,j}^{t} - 2C_{i,j}^{t} + C_{i+1,j}^{t}}{\Delta x^2} + E\frac{C_{i,j-1}^{t} - 2C_{i,j}^{t} + C_{i,j+1}^{t}}{\Delta y^2} \tag{15.4}$$

where Δt denotes the time interval, while i and j are the indices along the x and y coordinate axes, respectively. The x-axis is assumed parallel to the length of the river stretch and the y-axis is in the transverse direction in the cross section. Concerning expression (15.3), suitable average values are considered for the velocity components u and v, between the points in the water body, where the value of pollutant concentration is identified. Namely, the velocity components are considered longitudinally between two generic points $P(i,j)$ and $P(i-1,j)$ and transversely between two generic points $P(i,j)$ and $P(i,j-1)$.

At any point of the computational field $P(x,y)$, the water velocity components u and v and the pollutant concentration C are the average values along the vertical from the bottom to the free surface.

Following these assumptions, by means of the expressions (15.2), (15.3) and (15.4), the discretisation of Eq. (15.1) is written as

$$\frac{C_{i,j}^{t+1} - C_{i,j}^{t}}{\Delta t} + u\frac{C_{i,j}^{t} - C_{i-1,j}^{t}}{\Delta x} + v\frac{C_{i,j}^{t} - C_{i,j-1}^{t}}{\Delta y}$$
$$= E\frac{C_{i-1,j}^{t} - 2C_{i,j}^{t} + C_{i+1,j}^{t}}{\Delta x^2} + E\frac{C_{i,j-1}^{t} - 2C_{i,j}^{t} + C_{i,j+1}^{t}}{\Delta y^2} \qquad (15.5)$$

To reach realistic solutions, the general theory presupposes that the discrete Δt, Δx and Δy intervals comply with the constraint:

$$\Delta t \le \left[\frac{u}{\Delta x} + \frac{v}{\Delta y} + 2E\left(\frac{1}{\Delta x^2} + \frac{1}{\Delta y^2}\right)\right]^{-1} \qquad (15.6)$$

A further simplification also in explicit form can be introduced considering that in a river or stream the transverse component of velocity v is practically zero, while the longitudinal component varies according to the location of the point $P(x,y)$ in the cross section, as $u = u(i,j)$. This means that the transverse pollutant transport is realised only by dispersion and an application of this procedure is significant only when the dispersion coefficient E has a significant value.

Expression (15.5) can be rearranged leading to

$$C_{i,j}^{t+1} = \varepsilon_1 C_{i,j}^{t} + \varepsilon_2 C_{i-1,j}^{t} + \varepsilon_3 C_{i+1,j}^{t} + \varepsilon_4\left(C_{i,j-1}^{t} + C_{i,j+1}^{t}\right) \qquad (15.7)$$

in which

$$\varepsilon_1 = \Delta t\left[\frac{1}{\Delta t} - \frac{u}{\Delta x} - 2E\left(\frac{1}{\Delta x^2} + \frac{1}{\Delta y^2}\right)\right]$$

$$\varepsilon_2 = \Delta t\left(\frac{u}{\Delta x} + \frac{E}{\Delta x^2}\right)$$

Fig. 15.2 Water velocity
distribution at a cross section
of a river (two-dimensional
approach)

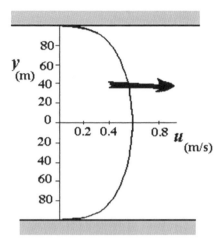

$$\varepsilon_3 = \frac{\Delta t E}{\Delta x^2}$$

$$\varepsilon_4 = \frac{\Delta t E}{\Delta y^2}$$

It is useful to illustrate the two-dimensional approach by means of two simple examples.

15.2   Examples

The water velocity in a river stretch is constant in the entire field and, in terms of mean value over the depth, varies in the horizontal x, y plane as shown in Fig. 15.2. A pollutant is continuously injected at the centre of the cross section identified by $x = 0$, with 100% concentration and infinite duration.

The pollutant is expected to be transported both longitudinally and transversely, eventually spreading all over the entire water body. The way the pollutant reaches a saturation value, after a certain time and in a downstream cross section, can be assessed by means of a two-dimensional model, such as that expressed by Eq. (15.7). The longitudinal velocity component u varies transversely as shown in the figure. The dispersion coefficient, $E = 5.0$ m^2/s, is constant in all the water body.

According to the geometric and hydraulic characteristics of the river, the space and time intervals, for the development of the model, are

$$\Delta x = 120.0 \text{ m} \quad \Delta y = 50.0 \text{ m} \quad \Delta t = 20 \text{ s}$$

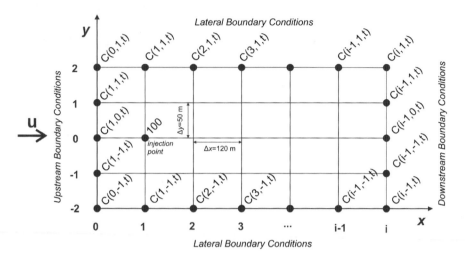

Fig. 15.3 Boundary conditions for the computational field

Together with the adopted values of u, they comply with the condition expressed by (15.6), to avoid the risk of unacceptable solutions. The pollutant is injected at point $P(0,0)$. At the initial time, concentration at all points is zero, except at the injection point.

As it is explained in the previous chapters, it is also necessary to determine the appropriate boundary conditions. All boundary conditions (upstream, downstream and lateral) are based on Neumann boundary conditions where

$$\frac{\partial C}{\partial x} = 0, \quad \frac{\partial C}{\partial y} = 0$$

The injection point should also be included as boundary condition with a continuous pollutant injection. The concentration value of injection is 100 (per cent) and represents the Dirichlet boundary condition.

In Fig. 15.3, further details are shown for the computational field and also for the boundary conditions in a classical non-staggered two-dimensional grid.

For the central axis, where $u = 0.60$ m/s, with the values indicated above, the coefficients of expression (15.7) are in Table 15.1, and for $j = 1$ (or $j = -1$ symmetrical), where the velocity flow is $u = 0.51$ m/s according to Fig. 15.2, the respective coefficients are in Table 15.2.

Solving Eq. (15.7) with these coefficients and with the appropriate values of the boundary nodes which were shown earlier results of pollutant concentration for various time instants are presented in Table 15.3.

In Figs. 15.4 and 15.5, the pollutant concentration variation is shown versus time in the central axis ($j = 0$) and at 50 m distance from the central axis ($j = 1$ and $j = -1$).

Table 15.1 Coefficients for Eq. (15.7) and for $j = 0$

$\epsilon_1 = 0.806$	$\epsilon_3 = 0.007$
$\epsilon_2 = 0.107$	$\epsilon_4 = 0.040$

Table 15.2 Coefficients for Eq. (15.7) and for $j = 1$ or $j = -1$

$\epsilon_1 = 0.821$	$\epsilon_3 = 0.007$
$\epsilon_2 = 0.092$	$\epsilon_4 = 0.040$

Table 15.3 Solution of Eq. (15.7) for the case examined

				$j = 0$ (0.0 m)					
$C(1,0,1)$	100.00	$C(1,0,2)$	100.00	$C(1,0,3)$	100.00	$C(1,0,4)$	100.00	$C(1,0,5)$	100.00
$C(2,0,1)$	10.69	$C(2,0,2)$	19.32	$C(2,0,3)$	26.34	$C(2,0,4)$	32.12	$C(2,0,5)$	36.94
$C(3,0,1)$	0.00	$C(3,0,2)$	1.14	$C(3,0,3)$	2.99	$C(3,0,4)$	5.24	$C(3,0,5)$	7.69
$C(4,0,1)$	0.00	$C(4,0,2)$	0.00	$C(4,0,3)$	0.12	$C(4,0,4)$	0.42	$C(4,0,5)$	0.90
$C(5,0,1)$	0.00	$C(5,0,2)$	0.00	$C(5,0,3)$	0.00	$C(5,0,4)$	0.01	$C(5,0,5)$	0.06
$C(6,0,1)$	0.00	$C(6,0,2)$	0.00	$C(6,0,4)$	0.00	$C(6,0,4)$	0.00	$C(6,0,5)$	0.00
				$j = 1$ (50.0 m) or $j = -1$ (−50.0 m)					
$C(1,1,1)$	4.00	$C(1,1,2)$	7.81	$C(1,1,3)$	11.45	$C(1,1,4)$	14.93	$C(1,1,5)$	18.26
$C(2,1,1)$	0.00	$C(2,1,2)$	0.80	$C(2,1,3)$	2.18	$C(2,1,4)$	3.98	$C(2,1,5)$	6.09
$C(3,1,1)$	0.00	$C(3,1,2)$	0.00	$C(3,1,3)$	0.12	$C(3,1,4)$	0.42	$C(3,1,5)$	0.94
$C(4,1,1)$	0.00	$C(4,1,2)$	0.00	$C(4,1,3)$	0.00	$C(4,1,4)$	0.02	$C(4,1,5)$	0.07
$C(5,1,1)$	0.00	$C(5,1,2)$	0.00	$C(5,1,3)$	0.00	$C(5,1,4)$	0.00	$C(5,1,5)$	0.00
$C(6,1,1)$	0.00	$C(6,1,2)$	0.00	$C(6,1,3)$	0.00	$C(6,1,4)$	0.00	$C(6,1,5)$	0.00

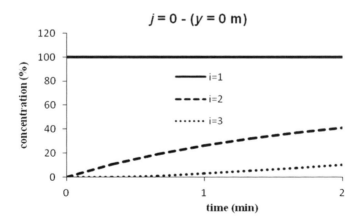

Fig. 15.4 Pollutant concentration along the central axis of the river

The pollutant, due to the continuous injection of infinite duration, is expected to spread over the whole river, reaching a concentration of 100% after few time steps.

It is also interesting to examine how the pollutant concentration varies streamwise and in the transverse direction against the axis of the stream.

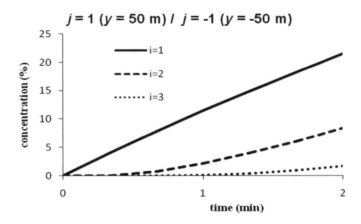

Fig. 15.5 Pollutant concentration at 50.0 m distance from the central axis

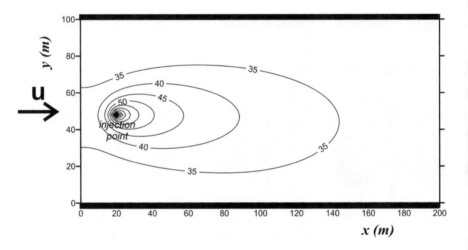

Fig. 15.6 Isoconcentration lines in the computational field for time instant $t = 0.25$ h

The computational grid of the example which is presented above has very few nodes for obvious reasons. A more illustrative example using the same numerical scheme (the upwind numerical scheme for 2D) and keeping the basic data unchanged is now presented. In this example, the spatial steps are lower.

Specifically let us consider a river with a width of 100 m and a constant flow velocity all over the computational field with $u = 1.0$ m/s. There is also a continuous injection point ($C = 100\%$) at the central axis 20 m from the upstream cross section of the river. Dispersion coefficient is $E = 5$ m²/s and the examined area is 200 m streamwise. The initial and boundary conditions are the same as in the previous example. As far as the discretisation is concerned, space step is 2 m for both dimensions (Δx, Δy) and time step is $\Delta t = 0.1$ s.

In the next Figs. 15.6, 15.7 and 15.8), the results of this example are presented as computed by the same numerical scheme.

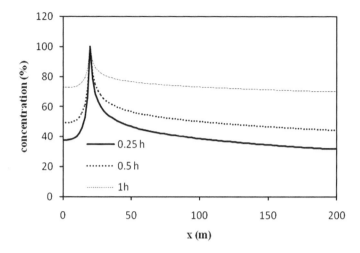

Fig. 15.7 Concentration profile at streamwise central axis

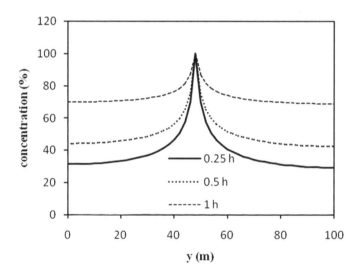

Fig. 15.8 Concentration profile at $x = 20$ m

It is obvious that the pollutant, due to the continuous injection (infinite duration), is also expected to spread over the whole river, reaching a concentration of 100% after several time steps.

15.3 An Outline of the 2D-Finite Element Method

The two-dimensional problem can be also solved by means of the finite element method. Some applications have been presented (e.g. Lee and Seo 2007) based on the fundamental theory already developed in the field of applied mathematics.

A very interesting procedure has been recently proposed by Rosati (2011, A two-dimension model for pollution transport in rivers, personal communication), which is an extension of the one described in Chap. 13 for the one-dimensional case.

The river flow and the pollutant concentration are assumed depth-averaged, like in the case examined in the preceding pages. They are, therefore, function only of the x and y coordinates and time t. The x, y plane is horizontal, with the x-axis parallel to the main direction of the flow. Moreover, the transverse velocity component, v, is null.

Like in the example described above, in all these procedures, the transverse velocity component is neglected and the transverse pollution transport is considered as caused only by dispersion. In such a context, Equation (15.1), in a more general form of nonpersistent pollutant with local injection or subtraction, is transformed into

$$\frac{\partial C}{\partial t} + u\frac{\partial C}{\partial x} = E\left(\frac{\partial^2 C}{\partial x^2} + \frac{\partial^2 C}{\partial y^2}\right) - kC \pm S \qquad (15.8)$$

As in Fig. 15.9, the plane x, y is divided into a grid of finite elements, which, for the sake of simplicity, are assumed square with dimensions $l \times l$. The river stretch is, therefore, simulated for the space

$$D = [0, L] \cdot [0, B]$$

Concerning the time variable, constant Δt intervals are considered.

The following conditions are assumed for Eq. (15.8):

$$C(x, y, 0) = \phi(x, y)$$
$$C(0, y, t) = \gamma$$

in which $\phi(x,y)$ is a function playing the same role as that in Chap. 13 for the one-dimensional case and γ is a constant value corresponding to the pollutant concentration $C(0,0,0)$. Moreover, another condition is imposed for the normal derivative of the concentration

$$\frac{\partial C(x, y, t)}{\partial v} = 0$$

at the frontier of the space D, namely, for

$$y = 0 \quad y = B \quad x = L$$

Introducing (according to the steps illustrated in Chap. 13) the function

Fig. 15.9 The grid of finite elements applied to the river stretch

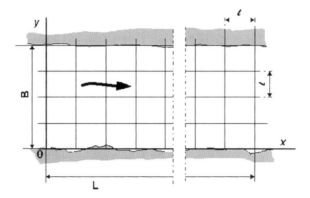

$$\Psi(x, y, t) = C(x, y, t) - \gamma$$

the expression (15.8) can be written as

$$\frac{\partial \Psi}{\partial t} + u \frac{\partial \Psi}{\partial x} = E \left(\frac{\partial^2 \Psi}{\partial x^2} + \frac{\partial^2 \Psi}{\partial y^2} \right) - k\Psi - k\gamma \pm S \tag{15.9}$$

With

$$\Psi(x, y, 0) = \phi(x, y) - \gamma$$
$$\Psi(0, y, t) = 0$$

$$\frac{\partial \Psi(x, y, t)}{\partial \nu} = 0$$

Assuming

$$l = u\Delta t$$

with a constant value of u for the entire river stretch, the numerical solution of (15.9) can be searched in a *variational* form, with a function $\phi(x, y)$ that minimises the functional

$$\iint_D \left\{ E\Delta t \left[(\Psi_x(x, y))^2 + (\Psi_y(x, y))^2 \right] + (1 + k\gamma\Delta t)\Psi^2(x, y) - \left[\Psi^{n-1}(x, y) - \Delta t(k\gamma \pm S) \right] \Psi(x, y) \right\} dxdy$$

The computation implies complicated procedures that are omitted in this description. The final expression, in explicit or matrix form, is not easy to be handled by means of a spreadsheet and requires more refined software.

As an overall consideration, the finite element method appears cumbersome and less manageable than the finite difference method. An advantage can be recognised considering that, in principle, the former is not restricted by the geometric characteristic of the element, which can assume whatever form and size in order to fit the river configuration.

Worthy mentioning is the application of the finite volume method, already described in Chap. 14 (Lin and Falconer 2005). Also for its applications, software packages are available (Falconer 2003; Wu and Falconer 1998).

The attempt to tackle the 3D approach is also interesting and some encouraging results are already obtained (Wu and Falconer 2000). Such an approach would permit the simulation of pollutant transport also along the vertical in the river cross section, giving a more complete picture of the behaviour of the injected pollutant. The presentation of the 3D approach is out of the scope of this book.

References

Ayyoubzadeh SA, Faramaz M, Mohammady K (2004) Estimating longitudinal dispersion coefficient in rivers. In: Proceedings of the 1st Asia-Oceania Geoscience Society, APHW session, Paper. 56-RCW-A69, pp 1–8

Blumberg AF, Ji ZG, Ziegler CK (1996) Modeling outfall plume behavior using a far field circulation model. J Hydraul Eng ASCE 122:610–616

Casulli V, Cheng RT (1992) Semi-implicit finite differences methods for three dimensional shallow water flow. Int J Numer Method Fluid 15:629–648

Casulli V, Stelling GS (1998) Numerical simulation of three-dimensional quasi-hydrostatic, free surface flows. J Hydraul Eng 124(6):678–686

Casulli V, Stelling GS (2011) Semi-implicit subgrid modelling of three-dimensional free-surface flows. Int J Numer Method Fluid 67:441–449

Chao X, Jia Y, Azad Hossain AKM (2009) 3D numerical modeling of flow and pollutant transport in a flooding area of 2008 US Midwest flood. In: Proceedings of the 33rd IAHR congress, Vancouver, Canada

Cheng RT, Casulli V, Milford SN (1984) Eulerian-Lagrangian solution of the convective-dispersion equation in natural coordinates. Water Resour Res 20(7):944–952

Dortch MS, Chapman RS, Abt SR (1992) Application of three-dimensional Lagrangian residual transport. J Hydraul Eng ASCE 118(6):831–848

Falconer K (2003) Fractal geometry: mathematical foundations and application. Wiley, Chichester/New York

Goltz MN, Roberts PV (1986) Three-dimensional solutions for solute transport in an infinite medium with mobile and immobile zones. Water Resour Res 22(7):1139–1146

Kalinowska MB, Rowinski (2009) Modified equation approach applied to 2D advection-diffusion equation with mixed derivatives. In: Proceedings of the 33rd IAHR congress, Vancouver, Canada

Lee ME, Seo W (2007) 2D finite element modeling of pollutant transport in tidal river. In: Proceedings of the 32nd IAHR congress, Venezia, Italy, vol 2, p 498 (printed summary)

Lin B, Falconer R (2005) Integrated 1-D and 2-D models for flow and water quality modelling. In: Proceedings of the 31st IAHR congress, Seoul, Korea, vol 1, pp 136–137 (printed summary)

Murillo J, Burguete L, Brufau PI, Garcia-Navarro P (2005) 1D and 2D models of solute transport using conservative semi-Lagrangian techniques. In: Proceedings of the 31st IAHR congress, Seoul, Korea, vol 1, p 507 (printed summary)

Murillo J, Garcia Navarro P, Brufau PI, Burguete L (2007) 2D numerical model for solute transport in shallow water flow over irregular geometries. In: Proceedings of the 32nd IAHR congress, Venezia, Italy, vol 1, p 199 (printed summary)

Wu Y, Falconer RA (1998) Two-dimensional ultimate quickest scheme for pollutant transport in free-surface flows. In: Third international conference on hydroscience and engineering, Cottbus, Germany, pp 1–13

Wu Y, Falconer RA (2000) A mass conservative 3-D numerical model for predicting solute fluxes in estuarine waters. Adv Water Resour 23:531–541

Wu Y, Falconer RA, Lin B (2001) Hydro-environmental modelling of heavy metal fluxes in an estuary. In: Proceedings of the XXIX IAHR congress, Beijing, China, Theme B5, pp 723–733

Chapter 16
Thermal Pollution

Abstract Water quality in rivers and streams can be altered also by an injection of heat that increases the water temperature, giving rise to thermal pollution. Water temperature is, therefore, the significant variable and can be used in the fundamental equation of pollutant transport, in the same way as the concentration of a chemical or biological pollutant. In the mathematical modelling of thermal pollution, the heat sources are key factors for water temperature simulation. Their evaluation requires proper investigation in the environment surrounding the water body.

16.1 The Discharge of Hot Water

As already mentioned in previous chapters, the water temperature is another important term to consider in water quality problems. An increased temperature in a water body alters the living conditions of aquatic weeds and animals, giving rise to the *thermal pollution*. With a density lesser than that of the receiving body, the hot water discharged into a river, lake or sea causes *density currents*, particularly noticeable on the free surface and in the upper layers of the water in a stream, where it gives rise to the *thermal plume* (Chevalier et al. 2007; Harleman and Stolzenbach 1972; Langford 1990). All over the river cross section, several levels of water at different temperature can be identified, with a vertical heat transport, normally from the free surface towards the river bottom. The density current modifies the original behaviour of the river and alters the natural circulation (Jurak and Winiewski 1989; Parker and Krenkel 1969; Zaric 1978).

From the overall viewpoint of physics, the thermal pollution is the worst type of natural water alteration, much worse than the chemical pollution caused by the injection of chemical wastes in the water body. It is indeed almost always possible to remove a chemical compound from a stream, before it reaches the environment, by means of a proper treatment process (the only constraints are the availability of suitable technology and the cost of the process implementation). Conversely, the

M. Benedini and G. Tsakiris, *Water Quality Modelling for Rivers and Streams*,
Water Science and Technology Library 70, DOI 10.1007/978-94-007-5509-3_16,
© Springer Science+Business Media Dordrecht 2013

removal of heat from water is governed by physical laws, and there is very little that man can do to prevent the heat to reach the environment (Parker and Krenkel 1969).

Thermal pollution is expected to increase in the near future owing to the growth of thermal power plants, which need huge amounts of water for cooling. Nevertheless, in some cases, hot water discharged from a cooling device can be drawn off for irrigation in the frozen zones of the Northern Hemisphere. In this case, it is not correct to speak about thermal pollution.

16.2   The Basic Equations

In a river stretch, assuming the water temperature, τ, as significant indicator of the thermal pollution, the fundamental equation of pollutant transport, in its three-dimensional form, is

$$\frac{\partial \tau}{\partial t} + \left(\frac{\partial v_x \tau}{\partial x} + \frac{\partial v_y \tau}{\partial y} + \frac{\partial v_z \tau}{\partial z} \right) = E \left(\frac{\partial^2 \tau}{\partial x^2} + \frac{\partial^2 \tau}{\partial y^2} + \frac{\partial^2 \tau}{\partial z^2} \right) \pm \Im \qquad (16.1)$$

where $\Im$ denotes the effect of heat exchange between the river and its environment (Harleman and Stolzenbach 1972).

Dealing with heat problems, besides the fundamental dimensions [MLT] (mass-length-time), it is convenient to consider also energy and temperature, which are here expressed, respectively, by [E] and [Θ]. Consequently, the dimension of $\Im$ is [ΘT^{-1}].

Concerning advection and diffusion, the temperature can be treated in the same way as the pollutant concentration dealt with in the preceding chapters, and all the expressed considerations are valid.

Unlike the chemical and biological pollution, in which the external contribution is normally located in few well-identified cross sections of the river, the exchange of temperature with the environment occurs along the entire river stretch. Consequently, term $\Im$ requires particular attention and must be considered at any step of the river length.

The heat exchange with the environment is caused by several phenomena, which will be briefly examined in the following paragraphs. For the purposes of analysis, the main aspects of a river, along with its length, some simplified assumptions can be made, but, for a more accurate investigation of the complex phenomena occurring between the water body and the surrounding environment, more detailed analyses are necessary, with the support of appropriate in situ measurements.

As already mentioned in Chap. 3, assuming that "the effect of the sum is the sum of single effects", the various phenomena can be examined separately from each other, with the assumption that $\Im$ is their global effect. In turn, also Eq. (16.1) is the resulting sum of advection, dispersion and $\Im$.

In the general approach, the heat transmission is assumed three-dimensional, as considered in the equation, where the temperature τ is a function of the x, y and z coordinates and time t

$$\tau = \tau(x, y, z; t)$$

This assumption allows for heat propagation in the river cross section to be examined. Particularly in the case of point source, the injected heat is initially concentrated in a restricted zone around the injection point and expands afterwards longitudinally and transversely, as it can be detected by the variation of temperature, which creates the already defined thermal plume. The stream temperature increases and the quantity of heat requires a long downstream reach in order to be extinguished. This entails an accurate analysis of the plume.

In a long river, it is practically more interesting to verify the effect of a heat injection on a long stretch. It is also interesting to know how the injected heat can be extinguished due to the various processes occurring in the water body. For this purpose, a one-dimensional approach is useful, assuming that the heat effect is expressed by the average temperature in the cross section. Therefore, Eq. (16.1) becomes

$$\frac{\partial \tau_x}{\partial t} + \frac{\partial v_x \tau_x}{\partial x} = E \frac{\partial^2 \tau_x}{\partial x^2} \pm \Im \qquad (16.2)$$

in which τ_x is the mean temperature on the river cross section, at the longitudinal x coordinate and time t.

The significance and the behaviour of the terms relevant to advection and dispersion are the same as those already described in Chap. 3. For this purpose, it is convenient to identify the elementary particle of the water body as in Fig. 16.1 extended on the entire depth and width of the stream; the volume of the elementary particle, V, is identified between the two successive cross sections at x and $x + dx$, with h being its average depth (Benedini 2011).

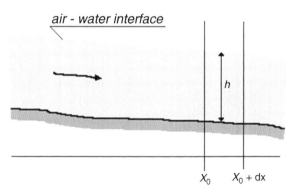

Fig. 16.1 The elementary particle in the one-dimensional approach

Normally an external concentrated man-made heat injection takes place only in a well-defined cross section of the river. Following the above assumptions, if x_0 is the location of such a cross section, the heat injection affects entirely the particle identified by x_0 and $x_0 + dx$, where, like in the other reaches of the river, there is also the continuous heat exchange between the water body and the environment.

16.3 Point Heat Injection

The most frequent case of man-made thermal pollution (Kinouchi et al. 2005; Miller 1984) is motivated by the assumption that the great quantity of running colder water in the stream can extinguish the heat, contributing to the environmental benefit (Zaric 1978). Consequently, the mass of the elementary particle undergoes the change of temperature

$$\Im = \frac{\partial \tau_x}{\partial t} = \frac{W_h}{\rho C_p V} \tag{16.3}$$

where

W_h = net heat injected in the river per unit time $[\text{E T}^{-1}]$
ρ = water density $[\text{ML}^{-3}]$
C_p = specific heat of water $[\text{E M}^{-1} \Theta^{-1}]$
V = volume of the elementary particle of water $[\text{L}^3]$

In the above dimensional expression, the energy, in technical units (metre-kilogram-second), is measured in *joules* (J); the temperature is measured in *centigrades* degrees (°C). Alternatively, the energy can be expressed in *kilocalories* (Cal). 1 Cal is equivalent to 4186.8 J.

The *specific heat* of water, C_p, varies with temperature, but for a river it can be assumed equal to 4187 J · kg^{-1} · °C^{-1} (joules per kilogram per centigrade), or about 1 Cal · kg^{-1} · °C^{-1}. Also the water density ρ varies with temperature and in a river is affected by the presence of substances dissolved or in suspension: for practical applications, it can be assumed $\rho = 1{,}000$ kg m^{-3}.

16.4 The Injection of a Hotter Flow

As illustrated in Fig. 16.2, a flow of hot water is injected in the river. This is the typical case of discharging the hot water from a thermal power station or from a chemical plant.

The heat coming into the elementary particle can be expressed as

$$W_h = \rho_0 q \, C_p \, \tau_0 \tag{16.4}$$

Fig. 16.2 Injection of hot
water discharged by a cooling
system

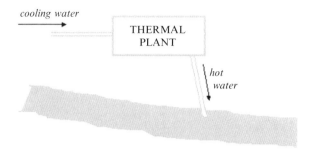

in which q is the flow of the injected hot water $[L^3 T^{-1}]$ and τ_0 its temperature.
In the adopted order of approximation, the hot water density, ρ_0, is lower than that
of the receiving body. For some high temperature, it can be around 980 kg m^{-3}.
The specific heat C_p can be assumed equal to that of the river water.

Expression (16.3) becomes, therefore,

$$\Im \cong 0.98 \frac{q\tau_0}{V} \tag{16.5}$$

with dimension $[\Theta \, T^{-1}]$.

The flow in the receiving body is increased by q, which in some cases can be of
the same order of that in the river. Consequently, the water velocity in the river can
increase.

16.5   Other Forms of Heat Injection

Besides the described point injection, which entails considerable flow of hot water,
several other forms of heat sources can be considered, not associated with an
increase of flow in the receiving body. In a simplified illustration, the heat injection
is realised in direct contact of the river water with a hotter body fully immersed in
the stream, identifiable as *heat exchanger* and shown in Fig. 16.3.

The heat enters the river by propagation through the casing of the exchanger that
is in direct contact with the water. In basic physics, this process is interpreted
according to the *Newton postulate* and expressed as

$$W_h = \eta(\tau_s - \tau_w)F \tag{16.6}$$

where

τ_s = temperature of the solid wall
τ_w = water temperature
F = area of the solid wall in contact with water

Fig. 16.3 Heat injection without increasing the water flow

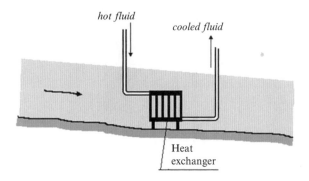

The *adduction coefficient* η [E T^{-1} L^{-2}Θ^{-1}] includes the effect of both heat convection and radiation occurring around the hot body and can be presented in the following form:

$$\eta = f + r \tag{16.7}$$

with f the *laminar propagation coefficient* and r the *radiation coefficient*. Suitable values of them are in the textbooks of thermodynamics. In practical cases of running water, f can be of the order of 200 Cal $\cdot$ m^{-2} $\cdot$ h^{-1} $\cdot$ °C^{-1} and slightly increases with water velocity and the solid wall temperature. Concerning r, a value of 4.0 Cal $\cdot$ m^{-2} $\cdot$ h^{-1} $\cdot$ °C^{-1} is usually acceptable.

16.6 Heat Exchange Between the River and Its Environment

This matter is somewhat complex, as many phenomena are involved, which in turn require measurements not always performed with the suitable precision (Chapra 1997).

The following paragraphs deal with some suggestions for a preliminary approach to the problem, in view of developing a predictive estimate of the river behaviour.

Particularly in large rivers, a considerable heat can be exchanged with the air, through the free surface of the stream. Significant can be also the exchange through the river bottom and the embankments, which can have a proper temperature due to geological and climatic conditions, varying according to the season and dependent on the latitude of the region. Another possible heat exchange is with the sediments transported by the flow and eventually settling at the bottom of the stream.

Although very important for the overall thermal balance, someone of these forms of heat exchange can be neglected for practical applications, although the scientific interest remains high.

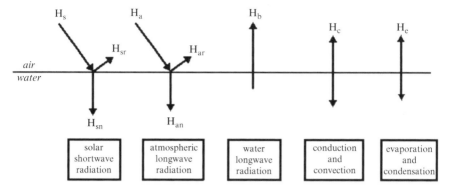

Fig. 16.4 The various processes of air-water heat exchange at the free surface of a stream

16.6.1 The Air-Water Heat Exchange

Several processes control the heat exchange between air and water, as illustrated in Fig. 16.4. Considering the elementary particle of the water body, the exchange occurs through its face in contact with the air and affects all the elementary volume as defined previously in the beginning of this chapter.

According to the figure, the heat sources, with dimension $[E\ L^{-2}\ T^{-1}]$, are identified as

H_s = heat due to the total short-wave solar radiation, part of which, H_{sr}, is reflected at the interface and absorbed by the atmosphere and the net part, H_{sn}, enters the water body

H_a = heat due to the atmospheric long-wave radiation, also consisting of a reflected part H_{ar} and a net part H_{an} entering the water body

H_b = outgoing heat due to long-wave back radiation

H_c = entering or outgoing heat due to the direct exchange with the atmosphere, by conduction and convection

H_e = heat lost due to evaporation

The net heat affecting the elementary particle, H_n, is given by

$$H_n = H_{sn} + H_{an} - H_b \pm H_c - H_e \tag{16.8}$$

The most usual values of these terms are given in Table 16.1. An accurate evaluation of these processes requires appropriate measurements and several calculations (Chapra 1997). The higher values in the table are for the days with intensive solar radiation, without clouds, and low humidity in the atmosphere.

The values refer to a long time interval (day); if considered in terms of second, they become relatively small.

In the adopted order of approximation, it is assumed that the heat passing through the free surface soon affects the entire mass of water in the stream.

Table 16.1 Range of
variability for the
air-water heat exchange
(Cal. m^{-2}. day^{-1})

	Min.	Max.
H_{sn}	980	7,100
H_{an}	6,300	8,400
H_b	6,500	10,000
H_e	400	8,200
H_c	-850	1,100

Consequently, expression (16.3) for the elementary particle of the river becomes

$$\Im' = \frac{H_n F_{f.s.}}{\rho C_p V} \tag{16.9}$$

where $F_{f.s.}$ [L^2] is the free surface area of the particle.

A more detailed evaluation of the heat exchange with the atmosphere should consider also the vertical propagation along the river depth, involving a three-dimensional analysis.

16.6.2   Heat Exchange with the River Bed

The heat exchange with the river bed occurs when a difference of temperature is clearly identified between the river water and the surrounding soil. The earth surface follows proper laws of absorbing and releasing the heat received from the sun. The atmospheric conditions, in terms of air temperature, humidity and precipitation, play an important role in the complex mechanism of air-soil heat exchange, in different way from the correspondent air-water heat exchange.

The phenomenon is, therefore, characterised by the persistence of climatic conditions and depends on the geographic location of the area taken into consideration.

According to Jurak and Winiewski (1989), in the Northern Hemisphere, the exchange is *positive* from the soil towards the water, in autumn and winter, and *negative* from the water towards the soil, in spring and summer, as sketched in Fig. 16.5. It depends also on the river depth becoming completely negligible for the deep streams. Some significant values are in Fig. 16.6 for the latitude of 40°. According to (16.3) the heat exchange is

$$\Im' = \frac{G F_{rb}}{\rho C_p V} \tag{16.10}$$

where

G = the heat passing through the unit area of the river bed in unit time [E L^{-2} T^{-1}]

F_{rb} = the area of the river bed in the elementary particle [L^2]

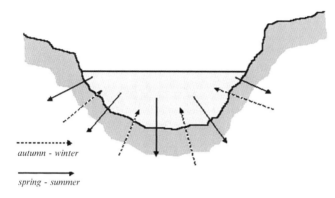

Fig. 16.5 Heat exchange between the river and its bed

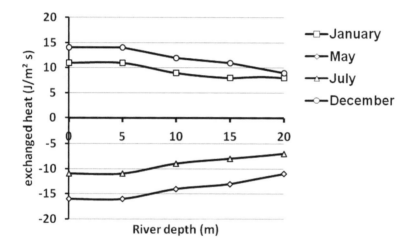

Fig. 16.6 Typical pattern of water and river bed heat exchange

In the elementary particle of the river taken into consideration for the model, the river bed includes the bottom and the embankments. For deep rivers, the contribution of the bottom is negligible.

16.6.3 Heat Exchange with Sediments

The sediments conveyed by the water can contribute to the thermal balance of a river. Normally, the sediments originate in the upstream reaches, where the ambient temperature is lower and the water velocity is higher. It is, therefore, expected that in the downstream reaches, the sediment temperature remains lower than that of the water, causing a subtraction of heat from the water volume.

The heat transfer involves all the sediments present in water, including those conveyed in the upper levels of the river cross section. The sediment presence is very difficult to analyse and formally interpret. Nevertheless, following also some experimental studies, there are good reasons to say that the predominant heat exchange is with the sediments settled on the river bottom, which give rise to a layer of considerable thickness, particularly in the most downstream reaches of the river.

Such an exchange can be expressed as

$$\Im = -\frac{H_{sed}}{\rho_{sed} \, C_{sed} \, h_{sed}} \qquad (16.11)$$

where

H_{sed} = heat flux between sediment and water per unit contact area [E L^{-2} T^{-1}]
ρ_{sed} = sediment density [ML^{-3}]
C_{sed} = specific heat of sediment [E M^{-1} Θ^{-1}]
h_{sed} = thickness of sediment layer [L]

In practice, not the whole sediment layer settled on the bottom contributes to the heat exchange with water, because there is also an exchange with the underneath original solid boundary, but in the order of approximation adopted, only the sediment-water heat exchange is considered. The relevant heat flux can be determined by the following empirical relationship:

$$H_{sed} = \rho_{sed} \frac{C_{sed} \, \alpha_{sed}}{h_{sign}} \left(\tau_{sed} - \tau_w\right) \qquad (16.12)$$

where

α_{sed} = sediment thermal diffusivity [L^{2}T^{-1}]
h_{sign} = significant thickness of the sediment layer [L]
τ_{sed} = sediment temperature [Θ]
τ_w = water temperature [Θ]

The thermal diffusivity is controlled by the mineralogical composition of the sediment and the status of aggregation of the solid grains. Its dimension is [L^{2}T^{-1}]. Table 16.2 gives some important values for the most frequent cases; being such values very small, the measuring units are conveniently adapted (in particular centimetre instead of metre).

The significant thickness h_{sign} is normally of the order of 0.10 m.

Appropriate average values are recommended, taking into account that in rivers with slow current, the sediment contains significant fractions of sand, gravel and stones. In the upper reaches of the river, the boulder and rocks substrate is

Table 16.2 Thermal properties of water and some typical materials

Type of material	α_s (cm²/s)	ρ (g/cm³)	C_p (cal/(g · °C))
Water	0.0014	1.000	0.999
Clay	0.0098	1.490	0.210
Soil, dry	0.0037	1.500	0.465
Sand	0.0047	1.520	0.190
Soil, wet	0.0045	1.810	0.525
Granite	0.0127	2.700	0.202

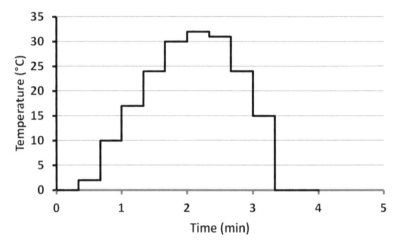

Fig. 16.7 Increase of the water temperature at the cross section due to heat injection

predominant. It is also worthy to point out that the sediment layer takes up very soon the water temperature and the heat exchange becomes, therefore, negligible.

16.7   An Example

The following example helps to better understand the outlines described in the preceding paragraphs.

A rectilinear channel with a constant and regular cross section receives a heat injection without flow at section $x = 0$, variable with time. Above the original value $\tau_0 = 18\ °C$, the water temperature in the cross section varies as shown in Fig. 16.7.

The heat propagation in the channel is simulated applying the finite difference method, according to the procedure illustrated in Chap. 12. The channel is divided into a sequence of six reaches of constant length $\Delta x = 25$ m; a sequence of 30 time intervals $\Delta t = 20$ s is considered. The average water velocity is $u = U = 0.30$ m/s for all the channel reaches, and the dispersion coefficient is $E = 5.0$ m²/s. The channel is 8.00 m wide and 0.90 m deep. A net heat $H_n = 2,000\ Cal \cdot m^{-2} \cdot day^{-1}$ is assumed.

The original water temperature in the channel is 18 °C and the heat propagation is considered above this original value. In the assumed conditions, the heat exchange with the environment is of a smaller order and, therefore, can be neglected. In fact, the value of $\Im$ would be of the order of 10^{-4} °C $\cdot$ s^{-1}, quite meaningless taking into account that in practice the water temperature can be measured at most with an approximation of 1 °C.

The fundamental equation becomes

$$-\frac{\tau(x,t)}{\Delta t} + \alpha_1\tau(x,t+1) + \alpha_2\tau(x+1,t+1) + \alpha_3\tau(x-1,t+1) = 0 \quad (16.13)$$

With the assumed values, the coefficients of the equation are

$$\alpha_1 = 0.066$$
$$\alpha_2 = 0.004$$
$$\alpha_3 = -0.020$$
$$-\frac{1}{\Delta t} = 0.05$$

Equation (16.13) gives rise to a linear system of six equations with six unknowns, to be solved applying the known rules of linear algebra.

For the time step $t = 0$, the system is

$-0.050\ \tau(1,0)$	+	$0.066\ \tau(1,1)$	+	$0.004\ \tau(2,1)$	−	$0.020\ \tau(0,1)$	=	0.000
$-0.050\ \tau(2,0)$	+	$0.066\ \tau(2,1)$	+	$0.004\ \tau(3,1)$	−	$0.020\ \tau(1,1)$	=	0.000
$-0.050\ \tau(3,0)$	+	$0.066\ \tau(3,1)$	+	$0.004\ \tau(4,1)$	−	$0.020\ \tau(2,1)$	=	0.000
$-0.050\ \tau(4,0)$	+	$0.066\ \tau(4,1)$	+	$0.004\ \tau(5,1)$	−	$0.020\ \tau(3,1)$	=	0.000
$-0.050\ \tau(5,0)$	+	$0.066\ \tau(5,1)$	+	$0.004\ \tau(6,1)$	−	$0.020\ \tau(4,1)$	=	0.000
$-0.050\ \tau(6,0)$	+	$0.070\ \tau(6,1)$				$0.020\ \tau(5,1)$	=	0.000

and, for any other time step, the right-hand side is to be adjusted taking into account the values of the preceding step.

After replacing the proper values of x and t, the solution is shown in Table 16.3, for some significant time steps. The temperature, in °C, is expressed by integer numbers because decimals are meaningless for practical applications.

The solution is shown also in Fig. 16.8, in which continuous curves replace the plotting in form of "discontinuous steps" always adopted in the preceding chapters. This graphical representation gives approximate values, but can be useful for a synthetic assessment of the heat propagation in the channel.

The figure shows how the thermal wave propagates in the channel and confirms the effect of advection and dispersion in the heat abatement.

Table 16.3 Significant values of the temperature in the channel

(min)	$x = 0$ (°C)	25 m (°C)	50 m (°C)	75 m (°C)	100 m (°C)	125 m (°C)	150 m (°C)
0	18	18	18	18	18	18	18
1	35	26	21	19	18	18	18
2	50	41	32	26	22	20	19
3	33	40	39	34	29	25	22
4	18	28	34	36	34	30	26
5	18	21	26	31	33	33	30
6	18	19	21	25	28	31	31
7	18	18	19	21	24	27	29
8	18	18	18	19	21	23	26
9	18	18	18	18	19	20	23
10	18	18	18	18	18	19	20

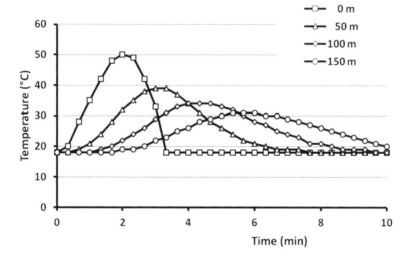

Fig. 16.8 Temperature profile of the channel, after the end of heat injection

References

Benedini M (2011) Water quality models for rivers and streams. State of the art and future perspectives. Eur Water 34:27–40

Chapra SC (1997) Surface water-quality modeling. WCB-McGraw Hill, Boston

Chevalier C, Hansen IS, Rassmussen EK, Vested HJ (2007) Modelling thermal plumes – Odense Fjord example. La Houille Blanche 5:29–36

Harleman DRF, Stolzenbach KD (1972) Fluid mechanics and heat disposal from power generation. Annu Rev Fluid Mech 4:7–32

Jurak D, Winiewski A (1989) Simple model of thermal pollution in rivers. W.P. –89-066, International Institute for Applied Systems Analysis, Laxenburg, p 24

Kinouchi T, Yagi H, Miyamoto M (2005) Long-term change of stream temperature: implication for impact of anthropogenic heat input due to urbanization. In: Proceedings of the 31st IAHR congress, Seoul, Korea, vol 2, pp 1008–1009 (printed summary)

Langford TEL (1990) Ecological effects of thermal discharges. Elsevier Applied Science, New York
Miller DS (1984) Thermal discharges: a guide to power and process plant cooling water discharged into rivers. British Hydromechanics Research Association, Kranfield, p 221
Parker FL, Krenkel PA (eds) (1969) Engineering aspects of thermal pollution. In: Proceedings, Vanderbilt University Press, Nashville, p 351
Zaric Z (1978) Thermal effluent disposal from power generation. Hemisphere Pub. Corp, Washington, DC

Chapter 17
Optimisation Models

Abstract Water quality control in rivers and streams is only a particular aspect related to the integrated water resources management. Simulation of water quality processes described in this book can assist in decision support procedures. Another type of models which can assist in decision making is the optimisation models. This chapter adds to the discussion on the use of optimisation models and particularly the linear programming models. The use of these models in relation to water quality simulation models is also discussed.

17.1 The Optimal River Management

The complexity of water problems requires more tools to focus on the role that a clean river plays in the context of water resources management and protection. An acceptable pollutant concentration in a river is the consequence of interventions involving all the contaminating sources. The reduction of the pollution load requests to control the urban discharges and the industrial and agricultural activities, together with the adoption of suitable treatment processes. All these initiatives have complex economic, social and political impact, which the *optimisation models* can help to understand. Optimisation models are, therefore, a powerful tool in the hand of the persons who have responsibility to intervene and decide how to utilise the available resources (Dorfman et al. 1972).

The optimisation models belong to a very large chapter of mathematics, described in numerous textbooks and experienced in many practical applications (Grant and Grant Ireson 1970; Loucks et al. 1981; Guggino et al. 1983). The following pages contain the description of some basic outlines of these models, focussing on their ties with the water quality simulation models described in the preceding chapters. The description is based on the assumption that the river is part of a *water resources system*, to be preserved in view of the benefits it can provide to the human community and to the conservation of the environment.

M. Benedini and G. Tsakiris, *Water Quality Modelling for Rivers and Streams*,
Water Science and Technology Library 70, DOI 10.1007/978-94-007-5509-3_17,
© Springer Science+Business Media Dordrecht 2013

Optimisation (or *programming*) models are tools able to identify, under a given set of conditions, an output value that can be considered, within a certain respect, the best of all the possible ones.

The substantial point of an optimisation model consists of an objective to be achieved through the system evolution and suitable to be translated into mathematical terms (*objective function*). The objective is chosen after discovering the main purpose for which the system has to be managed.

The objective function is formulated by means of the system variables and is accompanied by relationships stating the way in which the variables are interconnected in the problem (*constraints*).

The most general (*canonic*) formulation of a programming model is

$$
\begin{aligned}
\max \ & \mathbf{OF} \\
\textbf{subject to} \ & C_1 = \mathcal{E}_1 \\
& C_2 = \mathcal{E}_2 \\
& \quad \ldots \\
& C_m = \mathcal{E}_m
\end{aligned}
\tag{17.1}
$$

where the objective function **OF** and the constraints $C_1, C_2, \ldots C_m$ are functions of the $X_1, X_2, \ldots X_n$ variables, while $\mathcal{E}_1, \mathcal{E}_2, \ldots \mathcal{E}_m$ are pre-established values.

In the preceding statements, the optimisation is achieved through a maximum value of the objective function. According to the theory of programming models, the optimisation can be also achieved by means of a minimum value of the objective function and the formulation becomes

$$
\begin{aligned}
\min \ & \mathbf{OF} \\
\textbf{subject to} \ & C_1 = \mathcal{E}_1 \\
& C_2 = \mathcal{E}_2 \\
& \quad \ldots \\
& C_m = \mathcal{E}_m
\end{aligned}
\tag{17.2}
$$

Proper mathematical procedures are available for the development of the programming models, which are now currently applied to many complex problems of our life. It is worthy to mention that the optimisation models are effective tools for those situations where there is a *conflict* among the problem variables.

17.2 The Linear Programming Model

The objective function and the constraints can assume whatever form; an important type of programming models is characterised by having both the objective function and the constraints expressed as linear functions (or first degree polynomials) of the variables:

$$\max \mathbf{OF} = \max\{a_1.X_1 + a_2X_2 + \cdots + a_nX_n\} = \max \sum a_iX_i$$

subject to

$$C_1 = b_{1,1}X_1 + b_{1,2}X_2 + \cdots + b_{1,n}X_n = \sum_{b_{1,i}} X_i = \mathfrak{C}_1 \qquad (17.3)$$

$$\ldots$$

$$C_m = b_{m,1}X_1 + b_{m,2}X_2 + \cdots + b_{m,n}X_n. = \sum b_{m,i}X_i = \mathfrak{C}_m$$

where a_i and $b_{1,i}, b_{2,i}, \ldots b_{,m, i}$, for $i = 1, 2, \ldots n$, are constant values, either positive, negative or null, depending on the system definition.

This is the *linear programming model* (Dantzig 1963). Conversely, when linear formulations are not possible for the objective function and for the constraints, or when such formulations are better expressed in other forms than first degree polynomials, the procedure becomes non-linear, leading to *non-linear programming models*.

The linear programming models make up a very effective branch of applied mathematics, also in the water resources problems, and some descriptions will be given in the following paragraphs, with references to the principal textbooks of this subject.

17.3   Some Characteristics of the Linear Programming Models

In a linear programming formulation, the constraints can be expressed as inequalities and the right-hand sides are more properly *lower* (or *upper*) *bounds*:

$$\sum b_{j,i}X_i > \mathfrak{C}_j$$

for $j = 1, 2, \ldots m$

An inequality can be turned into equality by adopting appropriate *slack variables*: for instance, the inequality

$$\sum b_{j,i}X_i > \mathfrak{C}_j$$

is the equivalent to

$$\sum b_{j,i}X_i = \mathfrak{C}_1 - \hat{\imath}$$

where $\hat{\imath}$ is a nonnegative slack variable. In a general formulation, one or more X_i can be considered slack variable and their coefficients can assume only the values -1, $+1$ or 0.

The linear programming models can be applied only to phenomena suitable to be expressed in linear mathematical form. This is not always feasible. Several techniques are available for a linear interpretation of a non-linear behaviour, and the power of linear programming procedures allows assuring the validity of the "linearised" terms even though some approximation is necessary.

As seen above, the linear programming formulation leads to mathematical systems of linear equations, the terms of which can be conveniently put in a matrix form. All the rules of matrix computation theory can be, therefore, applied, and the computers have now available routines able to work out the matrix processes in a very easy manner.

17.4   An Example of Linear Programming Model

Advanced mathematical applications have made available tools able to deal with very complicated linear programming models, involving thousands of variables and terms, so that a practical limitation of the linear programming techniques is given only by the computing cost, which has become lower and lower following the computer technological progress.

To be acquainted with linear programming, a very simple example is presented below, leaving plenty of space for more thorough analysis of this matter with the aid of the specialised literature. The attention is focused on the way the model is formulated, considering the fundamental expressions presented in the preceding paragraphs.

A river stretch, like the one illustrated in Fig. 17.1, receives the discharge of a sewage network, which causes a pollution level in the downstream reaches and in the receiving coastal water. By means of a simulation model, the pollutant propagation is predicted based on suitable assumptions related to the flow conditions.

To restore a good environmental status, the construction of a wastewater treatment plant is proposed, adopting the appropriate technology, which involves a

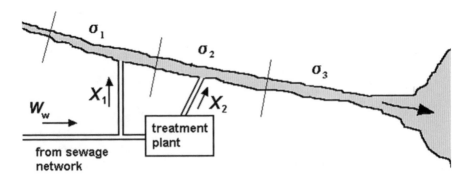

Fig. 17.1 A simple case for the application of linear programming model

proper cost. Anyhow, because a plant able to treat the whole amount of wastewater could be too costly, the possibility to discharge part of it directly into the river is considered, based on the assumption that the natural behaviour of the stream will reduce the pollution level in the downstream reaches.

With the notations of Fig. 17.1, the amount of water X_1 directly discharged and that X_2 treated by the plant are associated by the equation

$$X_1 + X_2 = W_w \tag{17.4}$$

in which W_w is the total amount of wastewater coming from the sewage network. Because such an amount is a predetermined constant, condition (17.4) means that if, for instance, X_1 increases, X_2 decreases, and vice versa. These two variables are, therefore, one another *conflicting*.

The discharge of untreated wastewater increases the pollution level in the river. Polluted water is a threat to the environment and makes the river water unsuitable for any use (and dangerous for life) in the downstream reaches. All these outcomes are eventually charged to the public authority responsible for the river management (Benedini 2002), with a cost S_1 that is proportional to the amount of wastewater directly discharged into the river:

$$S_1 = b_1 X_1 \tag{17.5}$$

with a coefficient b_1 that depends on the pollutant concentration $C_{\sigma 1}$ in the reach σ_1 and involves economic and social considerations, taking into account also the political aspects of the territory surrounding the river. On the other hand, the cost, S_2, of the wastewater treatment in the plant depends on the adopted technology and is a function of the pollution abatement provided by the involved physical, chemical and biological treatment. It is a function of the pollutant concentration acceptable in the stream, $C_{\sigma 2}$, in the river reach σ_2.

Taking all these considerations into account, the total cost of wastewater which is treated in the plant can be expressed as

$$S_2 = b_2 X_2 \tag{17.6}$$

with a convenient coefficient b_2. Both costs S_1 and S_2 are positive and are conditioned by the financial resources available by the responsible authority. Some upper limits, $\bar{S}_1$ and $\bar{S}_2$, respectively, are therefore to be considered.

Following the general considerations described in the preceding paragraphs, the linear programming model is, therefore, formulated as follows:

$$\min\ (b_1 X_1 + b_2 X_2) \tag{17.7a}$$

subject to

$$X_1 + X_2 \quad = \quad W_w$$

$$X_1 \qquad\qquad > \quad 0$$

$$X_2 \qquad\qquad > \quad 0 \tag{17.7b}$$

$$b_1 X_1 \qquad \leq \quad \mathcal{S}_1$$

$$b_2 X_2 \qquad \leq \quad \mathcal{S}_2$$

Expressions (17.7a) and (17.7b) give rise to a system of linear equations that can be solved using the available software packages. The solution gives the *optimal* value of the two variables X_1 and X_2, with the possibility of designing the treatment plant.

17.5   Post-optimal Analysis

The optimal solution given by the linear programming model is referred to the objective function that is structured as the summation of several terms. The value of the optimal solution obtained in this way does not refer to the single component terms, which can assume values even far from their individual optimum (this consideration is summarised in the statement that *the optimum of a sum is not necessarily the sum of the optima*). There is, thus, the need of a thorough examination of all the involved terms, in order to ascertain whether some components—at least the most important ones—can take up different values, without affecting the overall optimal solution. This is achieved through the *post-optimal analysis*.

Recalling the example of the previous paragraph, after modifying some terms of the problem, there is the possibility that X_1 and consequently X_2 have different values, remaining always in the in the optimal reach.

There are many ways of performing such an analysis (Duckstein and Plate 1987; Benedini et al. 1992; Benedini 1988); in most cases, it can be achieved by means of some computing procedures already inserted in the same algorithms of the linear programming techniques and currently available in the computer software. In these paragraphs, these procedures are shortly mentioned.

Relatively easy and immediate is the *dual analysis*, which allows determining how much a constraint, if supposed to be varied, can affect the overall optimal value of the objective function. More significant is the application of consolidated procedures developed in econometrics, which allow to point out how a variable can affect the others without altering the objective function.

A further step is the *sensitivity analysis*, which to some extent will be briefly described in the following chapters. Once the optimal value of the objective function is obtained, the attention is focused on how it is affected by the single variables.

The post-optimal analysis is a very useful technique to investigate the responsiveness of the system and to identify the components that require more attention.

17.6   Other Programming Models

There is plenty of space for the research aiming at producing new suitable tools for the optimisation procedures. Particularly in the field of water resources management, new cases are continuously examined, with innovative formulations and examples.

In this context, some directions can be mentioned, namely:

(a) *Integer* or 0/1 *linear programming* models, which are conventional linear programming models with the constraints suitable to take up integer or 0/1 values. These models are useful when decisions should be made on whether to follow a certain action or not (Tsakiris and Spiliotis 2011).
(b) Introduction of probabilistic aspects in the linear programming models, especially in the constraints (*chance constrained programming models*).
(c) Splitting of the procedure into sequential steps, each one to be optimised, putting into evidence their mutual interference (*dynamic programming*).
(d) Introduction of non-linear programming procedures.
(e) Use of *multicriteria* programming models where several conflicting criteria can be incorporated.

In practical applications, generally speaking, these tools are characterised by complex formulations and numerical calculations that can now be efficient, gaining from the progress in the computing facilities.

17.7   The Role of Programming Models

A common practice in water resources problems suggests that a single model cannot meet all the requirements relevant to the complexity of the case which is investigated. To perform a satisfactory analysis, several models are necessary, one another interacting and able to focus on the various aspects of the problem. Mutual relationships should be, therefore, envisaged among several models, of different nature, each one devoted to a particular aspect, with the aim to work out, as far as possible, a fully comprehensive analysis of the real problem which is examined.

In such a framework, the simulation models are able to analyse each single component of a more complex system. Besides the pollution transport in a water body, in water resources management, the simulation models deal with hydrology, river hydraulics and with the behaviour of reservoirs, channels, wastewater treatment plants and various water demands.

Fig. 17.2 Role and combination of several models for water resources management

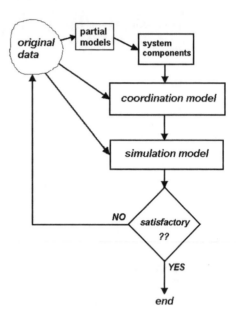

A programming model can also be utilised in order to identify the best geometries of the components, such as the size of reservoirs and the capacity of treatment plants. All these models can be considered *partial*.

More complex programming models (*coordination models*) deal with some overall features of the system, in order to find its operational rules according to physical and economic constraints. Once such programming models have provided the optimal operational rules, the entire system can be tested by means of a more comprehensive simulation model (Hufschmidt and Fiering 1966), which uses a very broad set of time series. Synthetic hydrology can be used for this purpose in order to build up a sequence of a convenient length, up to 100 or 1,000 years or more, having the same statistical characteristics as the recorded hydrological series. This enables one to find how many times the proposed rules can act in a satisfactory manner and to determine the probability of failures and wrong evaluations (Cicioni et al. 1981; Moncherino et al. 2007).

As shown Fig. 17.2, this model combination makes up an iterative process, to be carefully examined. It is worthy saying that this requires the best professional experience of the involved people.

If this probability of failures is judged too high, with the risk that the system is expected to act too many times in an unacceptable way, some components can be altered, adopting new operational rules and changing the size of the most important works. The described sequence of models can be then repeated until a satisfactory result is finally achieved.

References

Benedini M (1988) Water quality in river systems management: application of risk and reliability analysis. In: Ouazar D, Brebbia CA, Stout GE (eds) Computer methods and water resources: water quality, planning and management. Springer, Berlin, pp 237–250

Benedini M (2002) Water institutions in the new era and outstanding aspects of water problems. In: Tsakiris G (ed) Proceeding of the 5th international conference "Water resources in the era of transition", European Water Resources Association, Athens, Greece

Benedini M, Andah K, Harboe R (1992) Water resources management: modern decision techniques. A.A Balkema, Rotterdam

Cicioni GB, Giuliano G, Giulianelli M, Spaziani FM (1981) Models for the optimal management of water resources in the Tiber River basin. In: Proceedings of the 19th IAHR congress, New Delhi, India

Dantzig GB (1963) Linear programming and extensions. Princeton University Press, Princeton

Dorfman R, Jacoby HD, Thomas HA Jr (1972) Models for managing regional water quality. Harvard University Press, Cambridge

Duckstein L, Plate EJ (eds) (1987) Engineering reliability and risks in water resources, NATO ASI series no. 124. Nijhoff Publishers, The Hague

Grant EL, Grant Ireson W (1970) Principles of engineering economy. Ronald Press, New York

Guggino E, Rossi G, Hendricks D (eds) (1983) Operation of complex water systems, NATO ASI series no. 58. Nijhoff Publishers, The Hague

Hufschmidt MM, Fiering MB (1966) Simulation techniques for design of water resources systems. Harvard University Press, Cambridge

Loucks DP, Stedinger JR, Haith DA (1981) Water resources systems planning and analysis. Prentice Hall, Inc., Englewood Cliffs

Moncherino C, Palumbo A, Pianese D (2007) Mathematical modelling of contaminant propagation in natural and artificial free surface networks. In: Proceedings of the 32nd IAHR congress, vol 1, Venezia, Italy, p 273 (printed summary)

Tsakiris G, Spiliotis M (2011) Planning against long term water scarcity: a fuzzy multicriteria approach. Water Resour Manage 25(4):1103–1129

Chapter 18
Model Calibration and Verification

Abstract The model is always an interpretation of reality and is a valid tool only if it represents the reality correctly. The model calibration is, therefore, an essential step after its development. Calibration is performed by comparing the model output with the corresponding measured values. The verification is a further step for a more general evaluation of the model and requires a different set of measured data for testing the model performance. Combined steps of calibration and verification make up the model *validation*.

18.1 Calibration

A model, in any form, is always an interpretation of natural phenomena and is useful as long as it is capable of reproducing correctly the evolution of the reality, with the greatest possible accuracy. The model user has to be sure that his model is able to produce the real values successfully. The user is also conscious regarding the aspects in which the matching of model output and real measurements is weak or wrong.

As seen in the preceding pages, the development of a model requires the knowledge of the way the phenomenon evolves and needs some basic data, on which to adapt known formulations or, at least, to consider possible interrelations.

The set of data normally available for the construction of a model is limited to particular aspects of the reality. Similarly, the user familiarity with the phenomena, acquired in previous experience, does not cover necessarily the entire domain in which the problem has to be studied. Moreover, the necessity to comply with the formal rules imposed by the available mathematical procedures (or by practical tools, in case of physical models) imposes simplifications resulting in inadequate matching of the reality of the problem which is investigated. In other words, there is a possibility that the constructed model is not capable for producing a response entirely in line with the behaviour of the real phenomenon (Doneker et al. 2009; Kazakov et al. 2003; Mancini et al. 2000; Beck et al. 1993).

M. Benedini and G. Tsakiris, *Water Quality Modelling for Rivers and Streams*,
Water Science and Technology Library 70, DOI 10.1007/978-94-007-5509-3_18,
© Springer Science+Business Media Dordrecht 2013

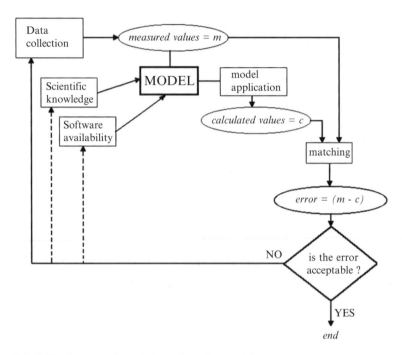

Fig. 18.1 Calibration procedure of the mathematical model

The model output is always a set of values of certain quantities, once a particular input is considered. The response of the model is correct if, given a certain input, the resulting output is confirmed by the corresponding real value acquired after the direct measurement related to the problem and in a way completely independent from the model itself. Following these considerations, after the model has been constructed, it is necessary to test it with the whole set of available data, proceeding to its *calibration*, as sketched in Fig. 18.1.

The core in the procedure consists of the *error*, the difference between the calculated and measured values of some significant terms. It has to be stressed that even the measured values may be inaccurate and, therefore, a thorough analysis of the accuracy of the available data is always necessary, as will be explained in Chap. 19.

Taking into account the overall complexity of the simulated mechanism, there is a very low probability that the error could be zero, even in ideal working conditions. It is, therefore, necessary to adopt a suitable *range of acceptability* of the error, with a pre-established threshold, which if exceeded the model cannot be accepted. This threshold is generally identified in the reach of the operator experience.

If, for a given range of acceptability, as illustrated in Fig. 18.1, the error can be accepted, the model is considered satisfactory. Vice versa, if the error exceeds the acceptable range, the model should be revised, modifying some components in a repeated iterative process, until an acceptable error is obtained. It is also worthy to point out that the model failures might lead to the conclusion that new experimental data are required. The model calibration becomes, therefore, another occasion for planning and carrying out proper campaigns of field data collection (Capodaglio et al. 2005).

The most immediate way to revise the model is to change the terms assumed as invariant, both for the entire process considered (namely, the constants) and within some pre-established conditions (the parameters). Generally, if all the assumptions are correct, the repetition of few runs is sufficient. In case the error remains unacceptable, it is necessary to revise also some basic assumptions or to check some basic mathematical formulations in the model.

Even though the model is satisfactorily calibrated, it must be born in mind that the model itself is always an interpretation of the reality valid only for the particular conditions on which the real problem is considered. These conditions are tied to the rest of the reality by means of constraints that could not be taken into a consideration at the very initial steps of the model development. Consequently, there is another set of factors that could cause the model to produce wrong or not rational results.

18.2   Verification

It is then necessary to test the model having in front, as far as possible, the entire reality in which the problem is placed, following a procedure known as *verification*. The final combination of calibration and verification steps is normally identified as *validation*. This entails a broader set of data, perhaps the comparison with other models and the commitment of expertise sometimes different from those employed for the construction of the model. The validation process should be thought as a philosophical exercise, by means of which the following items need appropriate attention:

(a) How and to what extent the available data can represent the conceptual interpretation of the considered phenomenon
(b) How and to what extent the adopted model can fit to such an interpretation
(c) How the available data can respond to the correct application of the adopted model

The process can benefit from some probabilistic procedures that will be better examined in the following chapters (Beven and Binley 1992).

The model validation is particularly related to the process of decision making, when its response is used to support an action related to the problem. The decisions are in the hands of people who should pay attention to several aspects of the problem that are not necessarily based on physical, biological or engineering expertise, but in economic, social and political considerations. Therefore, the results of a model application have to be inserted in the best possible frame of a

Fig. 18.2 Graphical
assessment of model
performance

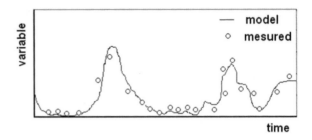

Fig. 18.3 Bar-chart
graphical assessment of
model performance

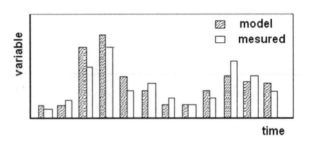

true interpretation of the reality and require to be presented in a way easy to be
understood by persons having various levels of experience and culture.

To conclude this discussion, there is another reason to stress that the mathematical model, like any other type of model, can be a successful tool only in the hands of
persons having high professional level and engineering judgement.

18.3   Quantitative Model Performance Assessment

The first somewhat subjective assessment of a model performance is to use graphical plots conveniently sketched with both results from measurements and from the
model used. Examples of these graphical methods are presented in Figs. 18.2 and
18.3. This visualisation method cannot produce quantitative criteria for the assessment of the model performance.

Therefore, some of the following statistical criteria can be used for comparing
the two time series, namely, the one produced by the model and the second from
measurements. These statistical criteria are considered objective and provide unbiased indicators of model performance. The most popular of these criteria are briefly
described below.

1. *Mean Absolute Error (MAE)*
   The MAE is a measure of the average deviation and is calculated as

$$\text{MAE} = \frac{\sum_{i=1}^{N} \left| Q_i^{\text{mod}} - Q_i^{\text{obs}} \right|}{N} \tag{18.1}$$

in which Q^{mod} is the i-th value produced by the model

Q^{obs} is the i-th value produced by measurement

N is the number of observations

It can be easily understood that the lower the value of MAE, the better the performance of the model.

2. *Bias Percentage (BP)*

The BP is defined according to the *BS criterion*

$$\text{BP} = \frac{\sum_{i=1}^{N} \left(Q_i^{\text{mod}} - Q_i^{\text{obs}} \right)}{\sum_{i=1}^{N} Q_i^{\text{obs}}} \times 100 \tag{18.2}$$

The BP criterion expressed as a percentage represents the average of residuals as a fraction of the average value. Obviously, the lower the BP, the better the performance.

3. *Root Mean Square Error (RMSE)*

The RMSE criterion is defined as

$$\text{RMSE} = \left[\frac{\sum_{i=1}^{N} \left(Q_i^{\text{mod}} - Q_i^{\text{obs}} \right)^2}{N} \right]^{\frac{1}{2}} \tag{18.3}$$

RMSE is a measure of the scatter of residuals. Values of RMSE near zero indicate good model performance.

4. *Relative Root Mean Square Error (RRMSE)*

The RRMSE is defined as

$$\text{RRMSE} = \frac{1}{\overline{Q}^{\text{obs}}} \left[\frac{\sum_{i=1}^{N} \left(Q_i^{\text{mod}} - Q_i^{\text{obs}} \right)^2}{N} \right]^{\frac{1}{2}} \tag{18.4}$$

in which $\overline{Q}^{\text{obs}}$ is the mean of observed values.

RRMSE is the same as RMSE, but normalised by the mean value of Q^{obs}, giving an indication of the scatter in relation to mean value.

5. *The Coefficient of Determination* (R^2)

R^2 expresses the strength of association between two data sets and is defined as

$$R^2 = \left[\frac{\sum_{i=1}^{N} \left(Q_i^{obs} - \overline{Q}^{obs} \right) \left(Q_i^{mod} - \overline{Q}^{mod} \right)}{\sqrt{\left[\sum_{i=1}^{N} \left(Q_i^{obs} - \overline{Q}^{obs} \right)^2 \right]} \sqrt{\left[\sum_{i=1}^{N} \left(Q_i^{mod} - \overline{Q}^{mod} \right)^2 \right]}} \right]^2 \qquad (18.5)$$

If R^2 approaches 1, a strong positive relationship between the two sets is observed. However, R^2 cannot guarantee that the model producing the first data set has high performance.

6. *Nash-Sutcliffe Coefficient of Efficiency (NSE)*

The NSE is defined as

$$\text{NSE} = 1 - \frac{\sum_{i=1}^{N} \left(Q_i^{mod} - Q_i^{obs} \right)^2}{\sum_{i=1}^{N} \left(Q_i^{obs} - \overline{Q}^{obs} \right)^2} \qquad (18.6)$$

The value of NSE varies from $-\infty$ to 1.0, with 1.0 being the optimal value. Values between 0 and 1 are generally considered acceptable, whereas negative values imply unacceptable model performance (Nash and Sutcliffe 1970).

There are many other numerical criteria, which have been used mainly for the assessment of hydrological models (Refsgaard and Knudsen 1996; Ahmed 2010) and have been applied to the water quality models only in some specific cases.

References

Ahmed F (2010) Numerical modeling of the Rideau valley watershed. Nat Hazards 55:65–84

Beck MB, Jakeman AJ, Mcalee M (1993) Construction and evaluation of models of environmental systems. In: Jakeman AJ, Beck MB, McAleer MJ (eds) Modelling change in environmental systems. Wiley, Cichester, pp 3–35

Beven KJ, Binley A (1992) The future of distributed models: model calibration and uncertainty prediction. Hydrol Process 6:279–298

Capodaglio AG, Boguniewicz J, Salerno F, Tartari G (2005) River and lake water quality simulation with GIS-integrated basin-scale models. In: Proceedings of the 31st IAHR congress, Seoul, Korea, 1, p 47 (printed summary)

Doneker RL, Sanders T, Ramachandran A, Thompson K, Opila F (2009) Integrated modeling and remote sensing systems for mixing zone water quality management. In: Proceedings of the 33rd IAHR congress, Vancouver, Canada

Kazakov B, Tzankov B, Lisev N, Bojkov V (2003) Kozloduy NPP water discharge- field study and modeling. In: Proceedings of the 30th IAHR congress, Thessaloniki, Greece, theme C, 1, pp 349–356

Mancini IM, Sole A, Boari G, Cavuoti C (2000) Calibration of a water quality model to simulate the effect of wastewater treatment plant in stream regime river. In: Proceedings of the international symposium on sanitary and environmental engineering, Trento, Italy

Nash IE, Sutcliffe IV (1970) River flow forecasting through conceptual models. J Hydrol 10:282–290

Refsgaard JC, Knudsen J (1996) Operational validation and intercomparison of different types of hydrological models. Water Resour Res 32:2189–2202

Chapter 19
Water Quality Measurements and Uncertainty

Abstract As already described in this book, in water quality modelling and in the sciences related to water resources, many decisions are based on the results of infield evaluations for the collection of the necessary data. These considerations are a matter for all kind of data used in the model development, and consequently, the issue of accuracy is very important for the validity of the model output. Inaccurate sampling and measurements may produce inaccurate results and wrong decisions for actions in water resources management. In this chapter, some basic aspects related to good practices in sampling and measurement of water quality indicators are presented, with particular relevance to the chemical compounds.

19.1 Methods of Analysis

The ability of gas to conduct heat, the capacity of a chemical solution to carry an electrical current or the ability of a coloured solution to absorb light can all be used as the basis for analytical methods able to detect the presence of a chemical compound or an element in water, as well as to measure its quantity (Sawyer et al. 1994). Any physical property of a compound or an element can be applied as the basis for an analytical measurement. The general principles of the main analytical methods applied in water analysis and examples of some current techniques are briefly presented. However, further information is contained in specialised books treating chemical and environmental analysis (e.g. Techniques, Valcarcel and Rios 1999; Rouessac and Rouessac 2007; Patnaik 2010).

19.1.1 Definite Methods

These methods are based on the laws that govern the chemical and physical parameters and concern the volume of titration reagent, the volume of titration

M. Benedini and G. Tsakiris, *Water Quality Modelling for Rivers and Streams*,
Water Science and Technology Library 70, DOI 10.1007/978-94-007-5509-3_19,
© Springer Science+Business Media Dordrecht 2013

generated product and the weight of test sample. The definite methods include volumetric analysis, gravimetry and potentiometry measurement, and also the activation analysis, the isotope dilution mass spectrometry, the voltammetry and polarography.

19.1.2   Relative Methods

These methods consist of comparing the sample to be analysed with a set of calibration samples of known content. The sample value is determined by interpolation of the measured quantity with respect to the *response curves* of the standard samples (ISO 1997; Quevauviller 2002). The requirements of relative methods include the matching of calibration sets, the elimination of interferences and the pretreatment of samples.

19.1.3   Comparative Methods

Comparative methods use a detection system particularly sensitive to the content of the molecules or elements to be determined in order to compare the sample to be analysed with a set of calibration samples (ISO 1997; Quevauviller 2002). X-ray fluorescence spectrometry is a typical example of a comparative method used for the analysis of liquid and solid samples.

19.1.4   On-line Monitoring Methods

According to a widespread opinion, on-line monitoring methods are a powerful tool for gaining information about the water quality. Sampling systems and sensors belong to the devices able to perform on-line monitoring.

Various sampling systems have been developed, which allow for a direct intake of surface water, while the collected sample is transferred to the measuring systems. The design of a monitoring system is obviously of great importance for collecting representative samples (Harmancioglu et al. 2003). Guidelines for designing a water quality monitoring network are provided by Harmancioglu et al. (1999). The filtration is the only pretreatment procedure required for surface water.

There are also *sensors* able to respond directly and rapidly to a variation of compound concentration. They are based on an *active microzone*, in which biological and/or chemical reactions may take place. An optical system may be connected to the microzone, able to respond in an accurate, continuous and direct manner to any variation of the quality indicator in the water body. Examples of sensors include biosensors and chemical sensors.

Sensors are generally easy to handle and use, although they present several disadvantages, such as the requirement of frequent calibration, motivated, in particular, by the growth of biofilm on their surface. Water quality indicators that can be monitored by sensors include, in particular, chloride, ammonia, nitrates, metal ions, cyanide, BOD, COD, TOC and pesticides.

19.2   Traceability

According to ISO (1994), traceability is defined as "the ability to verify the history, location, or application of an item by means of recorded identification", while, according to JCGM (2008), it is defined not only as a property of a measurement but also as a property of the reference standard. Needless to say, the problem of traceability is as old as the first quality measurements taken hundreds of years ago.

In general, traceability can be described as a logical and continuous care that keeps alive the effectiveness of any step of an analytical process. A comparison of measurements is significant only when they are expressed on the same scale or with the same units (Konieczka and Namiesnik 2009). According to these authors, the traceability of measurement depends, among others causes, on the proper functioning of measuring instruments, which can be assured by calibration using suitable comparison chemical compounds.

For a physical property, the measurement accuracy depends substantially on the quality of the measuring instruments, and, in principle, it does not depend on the object to be examined.

For a chemical compound, apart from the calibration of the gauging device, the measurement accuracy depends, to a significant degree, on the type of the sample and on how the analytical procedure is conducted. The necessity of a representative quantity of the material to be analysed is very important. Moreover, the notion of accuracy is difficult to define, and, consequently, the traceability is difficult.

The calibration of analytical instruments is not a significant source of problems for the chemical analysis, while the greatest problem is assuring the traceability of the entire analytical process. The most significant difficulties resulting from the sample preparation (before the measurement process itself), which are associated with the determination of traceability, are (a) the sample preparation, (b) the identification of the object to be measured, (c) the interference with other indicators, (d) the homogeneity of a sample, (d) the persistence of the sample and (e) the determination of uncertainty.

In practice (Bulska and Taylor 2003), the traceability for chemical measurements can be determined in by referring to an obtained value of reference standards (the reference values should come from laboratories with high international reputation) or by comparing an obtained value with reference measurements.

19.3 Uncertainty

19.3.1 Uncertainty of Sampling

From the viewpoint of statistics, a surface water sample belongs to a *population*, a statistical term that includes all the possible available measurements of a variable. It is impossible to record the whole population, since all the measurable terms could be measured at an infinite number of replicates and locations.

All the considerations on water quality are inferred by collecting just a few samples from a population. The sampling process has to collect data as accurate as possible, while the data, in order to be accurate, must be both *unbiased* and *precise*. Definition of these two terms is in following pages.

Generally, the random sampling eliminates almost completely the sampling bias, but new bias can be generated due to the errors of the analytical measurement. Consequently, the high variability in water quality indicators generates imprecise results.

In case that a sample is representative of the population from which it is collected, this sample may be used for extracting general conclusions concerning the entire population. This technique used in statistics is known as *statistical inference*.

Several factors are responsible of uncertainty during the sample analysis. An inappropriate methodology of measurements, an improper calibration of the instrument and a scarce representativity of the collected sample are just ones of the most remarkable and frequent causes. The uncertainty of estimating the *true* value of a quality indicator is related to the number of samples needed for a successful estimation.

The following paragraphs summarise some aspects of sampling for water quality problems. Specialised books on the subject can give additional information (Provost 1984; Prichard 1995; Quevauviller 2002). The measurement cases are presented below according to Prichard (1995).

19.3.2 Sampling Variance Significant-Measurement Variance Insignificant

The number of samples to be collected should be decided according to a sampling programme. The decision of how great can be the error on the final result plays an important role in the estimation of the size of the sample.

Provided that the uncertainty is known, the next step is to find the level of confidence. For example, a level of confidence of the 90% means that the 10% of the values of the samples lies outside the chosen uncertainty limits. In case the values are *normally distributed*, the limits of the sample arithmetic mean ($\bar{x}$) are

$$\bar{x} \pm \frac{t\,\widehat{\sigma}}{\sqrt{N}} \tag{19.1}$$

in which t is the value of the *Student probability distribution* at various "$1 - a$" *confidence levels* for $N - 1$ degrees of freedom, $\widehat{\sigma}$ is the sample standard deviation and N is the number of samples.

As known from statistics, the above expression holds provided that $N < 30$ and the standard deviation is not known but estimated from the sample standard deviation. It is easily deduced that the *true arithmetic mean* μ is, therefore, bounded as

$$\bar{x} - t_{N-1,\frac{a}{2}}\frac{\widehat{\sigma}}{\sqrt{N}} < \mu < \bar{x} + t_{N-1,\frac{a}{2}}\frac{\widehat{\sigma}}{\sqrt{N}} \tag{19.2}$$

The value of $t_{N-1,a/2}$ is obtained if N and $1 - a$ are known. For example, for $N = 10$ and level of confidence $1 - a = 95\%$, the tables of Student probability distribution give $t = 2.262$.

In case the values of samples are normally distributed with $N \geq 30$ and a sample arithmetic mean ($\bar{x}$), the value of the true arithmetic mean μ (at a confidence level of $1 - a$) falls within the boundaries

$$\bar{x} \pm \frac{Z_{a/2}\widehat{\sigma}}{\sqrt{N}} \tag{19.3}$$

in which $Z_{a/2}$ is the standardised variable of normal distribution for a confidence level $1 - a$.

The latter expression cannot be easily used because, according to the assumption, the standard deviation is known and the number of samples is great. However, for simplification, Equation (19.3) can be used for an estimation of the required number of samples if the sampling error (sampling uncertainty) is predetermined, the confidence level is given and a reliable estimate of the true standard deviation σ is secured. In this case, it can be written as

$$E = \bar{x} - \mu = Z_{a/2}\frac{\sigma}{\sqrt{N}} \tag{19.4}$$

or

$$N = \frac{Z_{a/2}^2\sigma^2}{E^2} \tag{19.5}$$

As an example, let be $1 - a = 95\%$, the standard deviation $\sigma = 0.30$ g/m^3 and allowed uncertainty ± 0.20 g/m^3, the required number of samples is

$$N = \frac{1.65^2 \times 0.30^2}{0.20^2} = 6$$

in which $Z_{a/2}$, taken from the tables of normal distribution, is 1.65.

19.3.3   Sampling Variance Insignificant-Measurement Variance Significant

In this case, a representative sample is required if a number of analyses (N_a) to be performed are

$$N_a = \frac{t^2 \sigma_m^2}{E_a^2} \tag{19.6}$$

in which t is the value relevant to the required confidence level, E_a is the total allowable uncertainty and σ_m^2 is the variance of measurements.

Again, if t of Student distribution may be replaced by $Z_{a/2}$, the number of necessary analyses is

$$N_a = \frac{Z_{a/2}^2 \sigma_m^2}{E_a^2} \tag{19.7}$$

that is, for $1 - a = 95\%$,

$$Z_{\alpha/2} = 1.65.$$

For practical applications, Equations (19.5) and (19.7) can be written, respectively,

$$N = 4\sigma^2 / E^2$$

and

$$N_\alpha = 4\sigma_m^2 / E_a^2$$

19.3.4   Sampling Variance Significant-Measurement Variance Significant

This question has no unique answer. According to Prichard (1995), the total uncertainty (E_{tot}) is given by combining both uncertainties in a unique expression that involves several terms not easily available in the practical application. This concept is left to more detailed analyses in scientific speculations.

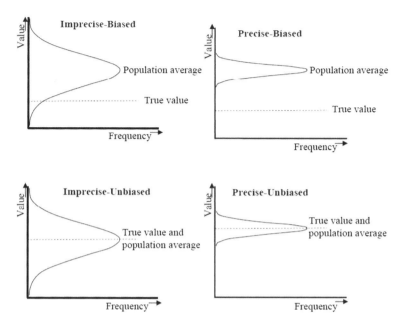

Fig. 19.1 Bias and precision (Redrawn from Prichard 1995)

19.3.5 *Uncertainty of Measurement*

The estimated uncertainty value is very important for the usefulness of a given analytical method.

Uncertainty of measurement is a nonnegative parameter characterising the dispersion of the values that could be attributed to a variable, based on the used information. *Definitional uncertainty* is the component of measurement uncertainty which results from a finite amount of detail in the definition of the variable. *Standard uncertainty* of a result of a measurement is expressed as a standard deviation (Konieczka and Namiesnik 2009).

The best operational tool to define the confidence level that may be attributed to a measurement is represented by the uncertainty. Moreover, uncertainty offers the advantage of being more explicit than vague terms such like bias, accuracy and precision.

Uncertainty and *measurement error* are two terms that are often prone to confusion. The measurement error is the difference between the expected and determined values, while uncertainty is a range into which the expected value may fall within a certain probability.

Bias of a measurement gives an estimate of how far the result is from the true value; *precision* gives information on the dispersion of the results (Fig. 19.1). The concepts of bias and precision in relation to accuracy are illustrated on the graphs shown in Fig. 19.2.

	Bias	Precision	Accuracy
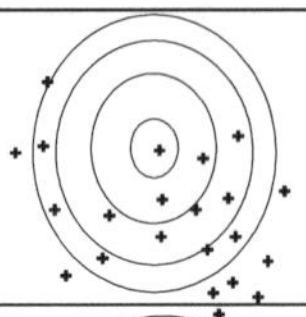	Biased	Imprecise	**Inaccurate**
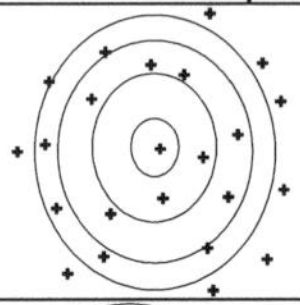	Unbiased	Imprecise	**Inaccurate**
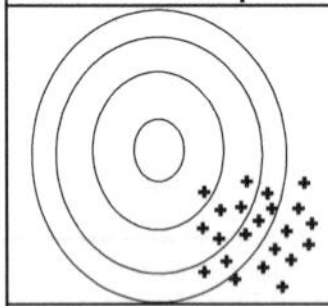	Biased	Precise	**Inaccurate**
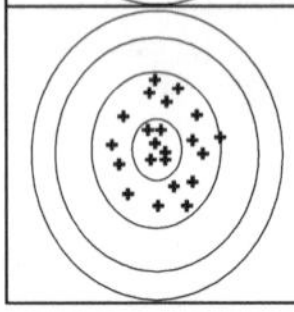	Unbiased	Precise	**Accurate**

Fig. 19.2 Examples of accurate and inaccurate measurements

According to Quevauviller (2002), examples of uncertainty sources for the determination of chemical compounds of water samples include (a) the matrix effects or interferences, (b) an incomplete extraction, (c) a contamination during sample pretreatment or collection, (d) an inaccurate certified value of reference materials, (e) imperfections of the measurement method or of instrumentation and (f) insufficient definition of the variable to be determined.

The interval, in which the value of a measurement may be expected, with a defined confidence level, is represented by the uncertainty, which includes both the components related to systematic and random effects (Quevauviller 2002).

19.3.6 Estimation of Total Uncertainty

The combined uncertainty value (S_{total}) is calculated by using the following expression (Quevauviller 2002):

$$u_{total} = \sqrt{u_{sampling}^2 + u_{measure}^2 + u_{pop}^2} \qquad (19.8)$$

where

$u_{sampling}^2$ = is determined by repeated collection of at least seven (identical) samples

$u_{measure}^2$ = is determined by repeated analyses (in different analysis batches) of homogeneous samples

u_{pop}^2 = is related to the sample population and is more critical for analysis since it conditions the variability of the indicator concentrations

The number of the samples to be collected and analysed in order to achieve a given uncertainty is influenced by the magnitude of each of the components of Eq. (19.8).

19.4   Quality Assurance and Quality Control

Related to the uncertainty are two important items defined by the ISO, namely, the *quality assurance* (QA) and the *quality control* (QC).

Quality assurance is defined as all those planned and systematic actions necessary to provide adequate confidence that a product or service will satisfy with given requirements for quality. There are five independent elements which are the pillars for quality assurance of an analytical measurement (Konieczka and Namiesnik 2009): (a) the use of Certified Reference Materials, (b) the validation of the applied analytical procedures, (c) the evaluation of uncertainty, (d) the assurance of measuring traceability of the obtained results and (e) the participation in various inter-laboratory comparisons.

On the other hand, quality control is defined as the operational techniques which are used to fulfil the requirements of quality. Quality control would include the following operational techniques: (a) the determination of precision by analysis of replicate samples, (b) the detection of impurities in the reagent or interferences by analysing blank samples and (c) a check of the accuracy of the detection method by analysing reference materials.

Traceability and uncertainty are the basic terms characterising an analytical result. Figure 19.3 presents a schematic representation of this concept.

Every "producer" of analytical results should bear in mind that it is necessary to present the chemical analysis result together with these two fundamental qualities.

Fig. 19.3 Uncertainty and traceability for producing a reliable analytical result (*SI* system international) (Konieczka and Namiesnik 2009)

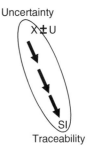

19.5 Biological Indicators

The considerations developed in the preceding paragraphs concerning the chemical and physical indicators hold for all the quality problems related to the environmental preservation and can be applied considering the quality indicators relevant to the vital aspects of aquatic animal and plants. As anticipated in Chap. 7, such indicators are suitable to be measured and provide data for the development of water quality models.

It is worthy to mention that the aquatic life is strictly related to the presence of chemical compounds and physical entities for which particular limits are requested. The scientific literature contains abundant description with reference to the living conditions of all the species that characterise the river environment, and, consequently, the relevant indicators and corresponding limits can be taken into consideration in the model application (Champ et al. 2009; Schmutz et al. 2007; Aguiar et al. 2011; Allan 1995).

Nevertheless, there are some biological indicators that can be directly considered with their concentration in water, in the same way as the indicators mentioned in the preceding pages. They are principally the bacteria and the viruses, for which the concentration value to be introduced in the model requests accurate analyses with appropriate statistical assessment.

19.6 Sediment and Suspended Solids

Water quality in rivers and streams can be altered by solids grains, removed from the soil and presented in water in the form of suspension. They belong to the topic of sediment and solid transportation, for which conspicuous researches have been conducted and are actually in progress (Bartram and Balance 1996; Gray et al. 2000; Ritchie and Cooper 1988; Alexandrov et al. 2003). As already mentioned in this book, sediments can contain elements and chemical compounds suitable to interact with water and are very often the cause of high pollution level. Moreover, high concentration of sediments is itself a form of water quality alteration, which can be appreciated by an increased turbidity that makes the water unsuitable for many uses. Sediments are also responsible for the change of river morphology, through the alternate mechanisms of deposition and erosion.

Specific models are available for the simulation of sediment transport in rivers and streams, developed on fundamentals not strictly in line with the basic scientific interpretations on which this book is based. Anyhow, the sediment concentration can be used as a quality indicator for several cases of water pollution (Chapra 1997), in a way that is similar to that of chemical compounds and elements considered in the preceding paragraphs.

Sediment concentration is determined by means of samples taken from the water body, which undergo appropriate analyses directly in the field and in laboratory. The results of such analyses request all the considerations regarding the accuracy of the data used as the input for described models.

19.7 Physical Measurements

The water quality indicators considered in the various chapters of this book include some physical terms to be handled in the same way as described above for the development of the water quality models. Among these, the water temperature is fundamental not only for the simulation of thermal pollution as described in Chap. 16 but also because it is determinant for the behaviour of some chemical and other physical parameters (Hannah et al. 2007; Poole and Berman 2001; Brown et al. 2006; Dodds 2002; Deas and Lowney 2000), as also mentioned in Chap. 8.

Water temperature in rivers and streams is evaluated by direct in situ measurement, but the complex reality of a water body greatly affects the accuracy of the obtained data. To provide reliable and significant values, the results of the measurements have to be carefully analysed, and an application of the various methods mentioned in this chapter is necessary, particularly those relevant to the statistical evaluations.

19.8 Hydrological Data

Proper hydrological and hydraulic information is necessary for the development a consistent water quality model. Not only the water velocity and the water level are requested for the correct simulation of the pollutant transport, but the knowledge of some meteorological and climate data can be useful to define the specific environmental conditions to which the model is pertinent.

Collection and analysis of such data is a matter of specific procedures relevant to the various chapters of hydrology, and the reader can refer to the numerous textbooks dealing with the subject. The accuracy of direct measurements in the water body relies to a large extent on the instruments to be utilised, for which the technological progress is continuously providing improved and reliable equipments.

The abundant literature available in this scientific field provides also the methods for the proper statistical interpretation of the measurements in order to assess the model validity (Herschy 1999; Gierke 2002).

References

Aguiar FC, Feio MJ, Ferreira MT (2011) Choosing the best method for stream bioassessment using macrophyte communities: indices and predictive models. Ecol Indic 11(2):379–388
Alexandrov Y, Laronn JB, Ian Reid I (2003) Suspended sediment concentration and its variation with water discharge in a dryland ephemeral channel, northern Negev, Israel. J Arid Environ 53(1):73–84

Allan J (1995) Stream ecology: structure and function of running waters. Chapman & Hall, London

Bartram J, Balance R, United Nations Environment Programme, World Health Organization (eds) (1996) Water quality monitoring – a practical guide to the design and implementation of freshwater quality studies and monitoring programmes. E & FN Spon, London/New York. ISBN 0 419 22320 7 (Hbk) 0 419 21730 4 (Pbk)

Brown LE, Hannah DM, Milner AM (2006) Thermal variability and stream flow permanency in an alpine river system. River Res Appl 22:493–501

Bulska E, Taylor PH (2003) On the importance of metrology in chemistry. In: Namieśnik J, Chrzanowski W, Żmijewska P (eds) New horizons and challenges in environmental analysis and monitoring. CEEAM, Gdańsk

Champ WST, Kelly FL, King JJ (2009) The water framework directive: using fish as a management tool. Biol Environ Proc R Ir Acad 109b(3):191–206

Chapra SC (1997) Surface water-quality modeling. WCB-McGraw Hill, Boston

Deas ML, Lowney CL (2000) Water temperature modeling review-central valley. California Water and Environmental Modeling Forum, Sacramento

Dodds WK (2002) Freshwater ecology concepts and environmental applications. San Diego Academic Press, San Diego

Gierke JS (2002) Measuring river discharge. ENG5300 engineering applications in the earth sciences, Michigan Technological University, Houghton

Gray JR, Douglas Glysson G, Turcios L M, Schwarz GE (2000) Comparability of suspended-sediment concentration and total suspended solids data. Water-resources investigations, U.S. Department of the Interior, U.S. Geological Survey, WRIR 004191

Hannah DM, Brown LE, Milner AM, Gurnell AM, McGregor GR, Petts GE, Smith BPG, Snook DL (2007) Integrating climate-hydrology-ecology for alpine river systems. Aquat Conser Mar Freshw Ecosyst 17:636–656

Harmancioglu NB, Fistikoglu O, Ozkul S, Singh VP, Alpaslan N (1999) Water quality monitoring network design, Water science and technology library. Kluwer Academic Publishers, Dordrecht/Boston

Harmancioglu NB, Ozkul S, Fistikoglu O, Geeders P (2003) Integrated technologies for environment monitoring and information production. NATO science series. IV, Earth and environmental sciences, 23

Herschy RW (1999) Hydrometry: principles and practices. Wiley, Chichester

ISO (1994) Quality management and quality assurance vocabulary, ISO 8402. International Organization for Standardization, Geneva

ISO (1997) Calibration of chemical analyses and use of certified reference materials, ISO guide 32. International Standardization Organization, Geneva

JCGM (2008) International vocabulary of metrology, basic and general concepts and associated terms (VIM), Joint Committee for Guides in Metrology, Geneva, Switzerland

Konieczka P, Namiesnik J (2009) Quality assurance and quality control in the analytical chemical laboratory: a practical approach. CRC Press/Taylors & Francis Group, Boca Raton

Patnaik P (2010) Handbook of environmental analysis: chemical pollutants in air, water, soil and solid wastes, 2nd edn. CRC Press/Taylor & Francis Group, Boca Raton

Poole GC, Berman CH (2001) An ecological perspective on in-stream temperature: natural heat dynamics and mechanisms of human-caused thermal degradation. Environ Manage 27:787–802

Prichard E (Co-ordinating Author) (1995) Quality in the analytical chemistry laboratory, ACOL series, Wiley, Chichester

Provost LP (1984) Statistical methods in environmental sampling. In: Schweitzer GE, Santolucito JA (eds) Environmental sampling for hazardous wastes, ACS symposium series 267. American Chemical Society, Washington, DC: 79–96

Quevauviller P (2002) Quality assurance for water analysis. Wiley, Chichester

Ritchie JC, Cooper CM (1988) Comparison of measured suspended sediment concentrations with suspended sediment concentrations estimated from Landsat MSS data. Int J Remote Sens 9(3):379–387

Rouessac F, Rouessac A (2007) Chemical Analysis: modern instrumentation methods and techniques, 2nd ed., Wiley, Bognor Regis, p 600

Sawyer CN, McCarty P, Parkin GF (1994) Chemistry for environmental engineering, 4th edn, Civil engineering series. McGraw-Hill International Editions, New York

Schmutz S, Cowx IG, Haidvogl G (2007) Fish-based methods for assessing European running waters: a synthesis. Fisheries Manage Ecol 14(6):369–380

Techniques, 2nd edn. Wiley, Chichester, Geneva, Switzerland

Valcarcel M, Rios A (1999) Traceability in chemical measurements for the end users. Trends Anal Chem 18:570–576

Chapter 20
Model Reliability

Abstract Model calibration and validation are not a sufficient step to completely ascertain its reliability, because there are always some reasons of incertitude due to the variability of the terms on which the model is developed. The choice of the parameters, the imprecision of input data and the inner mechanisms of the mathematical procedure may give an unreliable solution. To overcome such a disadvantage, sensitivity and uncertainty analysis can be of assistance. They benefit from basic concepts of statistics, able to focus, in particular, on the discrepancies between the model output and the corresponding values measured in the field. Complex procedures are available, the fundamentals of which are briefly described in this chapter.

20.1 The Effects of Parameter Variability

Even though the model has been successfully calibrated and validated, a doubt still remains about its capability of producing reliable results when it is currently applied to the largest set of practical cases. Any application is conditioned by the value of some terms that normally cannot be measured correctly or are assumed *a priori*, starting from exogenous considerations. This concerns principally the parameters, which play the most important role in the model behaviour. In other words, this puts the question: "How much the model results can vary if some parameters are chosen out of a predetermined set of values?"

Such a question leads to *sensitivity analysis*, by means of which the model is run repeatedly after assuming several values for the parameters incorporated in its internal structure (Breierova and Choudhari 2001). The analysis can be explained with the illustration of Fig. 20.1.

For a given input X, the model produces an output Y that depends on the specific value assumed for its parameters, each one characterised by its own range of variability. The output, consequently, has also its variability range, which, moreover, depends on the inner structure of the mathematical formulations. It may

Fig. 20.1 The effect of parameters variability

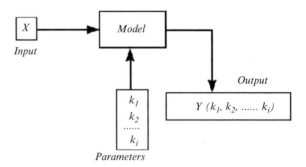

Input

Parameters

Output

Table 20.1 Initial values and consequent output

	Unit	Value
v	m/s	0.1
E	m^2/s	5.0
k	s^{-1}	0.0005
$C(100,1)$	%	0.011
$C(100,3)$	%	4.34
$C(100,5)$	%	14.89

happen that parameters having a limited range of variability cause an output with a large range of values: the model is, therefore, highly sensitive. In practice, this means that great attention must be given in the model application, and an improper choice of a parameter could give unreliable results. Vice versa, if parameters with a great variability range give an output with a narrow range, the model is "good" and can be applied with high reliability.

Obviously, the amplitude of the variability ranges is chosen arbitrarily, according to the operator's experience and following the specific evaluations of the reality in which the model is developed.

There are several ways of conducting the sensitivity analysis (Saltelli et al. 2008). Generally speaking, all the ways require repeated model runs assuming well-defined values of one or several of the most significant parameters.

The simplest and most immediate procedure consists of comparing the results of several model runs after assuming the largest set of parameter values, compatible with the real aspects of the river and the pollutant discharge. The procedure is shown by the following example.

The analytical model described in Chap. 9 is applied to a river stretch having the characteristics shown in Table 20.1; at time $t = 0$, a continuous source of pollution starts. The model output $C(x,t)$ is considered in terms of pollutant concentration at a cross section $x = 100$ m downstream from the injection point and at $t = 1$ min, $t = 3$ min and $t = 5$ min, giving the concentrations $C(100,1)$, $C(100,3)$ and $C(100,5)$, respectively. For the model application, the water velocity v, the dispersion coefficient E and the reaction coefficient k are assumed after considerations that do not depend on the model itself and are kept constant during each model run: they are, therefore, the parameters to be considered for their influence on the model output.

The model sensitivity can be determined separately for each parameter. For the first attempt, keeping v and k constant, the dispersion coefficient has been varied in a large interval, as shown in Table 20.2 with the corresponding output. The same procedure is adopted to analyse the effect of v and k. The output values after the parameter variation (*perturbation*) are compared with the "initial" values of Table 20.1. This procedure is called *one-factor-at-a-time* and conventionally abbreviated as OAT.

At a first glance, Table 20.2 shows that the parameters have a different effect on the model. For instance, at $t = 5$ min, an 8-time perturbation of k reduces the output only to the half, while the same variation of E increases the output more than 38 times. Moreover, the sensitivity depends on the time: for instance, an 8-time perturbation of v causes at 1 min an output variation of 22 times, while at 5 min, the variation is only 7 times. This also confirms that the effect of the parameter perturbation is much greater during the transition phase of the pollutant transport in the river, while it decreases as soon as the saturation is approached.

The effect of the dispersion coefficient E is paramount, particularly at the beginning of the pollutant injection. In practice, all these considerations suggest that the model must be applied at proper time and space and its terms must be assumed after a thorough examination of the real configuration to be modelled.

It is worthy to notice that the above example is based on an application of the analytical model: similar considerations can be performed for a numerical model, taking into account that the discretisation and the other approximations in the calculation can further increase the cause of output alteration.

20.2 A More Refined Analysis

The procedure illustrated in the previous paragraph can give an immediate but limited answer, conditioned by the way the parameters vary. Following are the fundamental aspects of a more comprehensive procedure, the *first-order sensitivity analysis* (Chapra 1997; Chapra and Canale 1988; Yen et al. 1986).

After several runs of the model with different values of the parameter p, each one giving the corresponding output C, the couples $\{p_i, C_i\}$ allow a function $C = C(p)$ to be determined, either through a statistical regression or in a graphical form.

In the general context of the reality to which the model is applied, the derivative

$$\underline{C}' = \frac{\partial C}{\partial p} \tag{20.1}$$

can be considered as an analytical expression of the model sensitivity with respect to the parameter p. The behaviour of $\underline{C}'$ shows how the model is sensitive, both in qualitative and quantitative way.

Table 20.2 Model output for various values of the significant parameters

Sensitivity to E
$k = 0.0005$ s^{-1}
$v = 0.1$ m/s

E (m²/s)	1 min C(100,1) (%)	3 min C(100,3) (%)	5 min C(100,5) (%)
1.25	0.000	0.009	0.733
2.50	0.000	0.501	5.040
5.00	0.012	4.344	14.894
10.00	0.618	14.359	28.313

Sensitivity to v
$k = 0.0005$ s^{-1}
$E = 5.0$ m²/s

v (m/s)	1 min C(100,1) (%)	3 min C(100,3) (%)	5 min C(100,5) (%)
0.05	0.007	2.779	9.786
0.1	0.011	4.344	14.894
0.2	0.029	9.554	29.530
0.4	0.153	30.798	66.464

Sensitivity to k
$v = 0.1$ m/s
$E = 5.0$ m²/s

k (s^{-1})	1 min C(100,1) (%)	3 min C(100,3) (%)	5 min C(100,5) (%)
0.0005	0.011	4.344	14.894
0.001	0.011	4.047	13.397
0.002	0.011	3.515	10.863
0.004	0.009	2.657	7.213

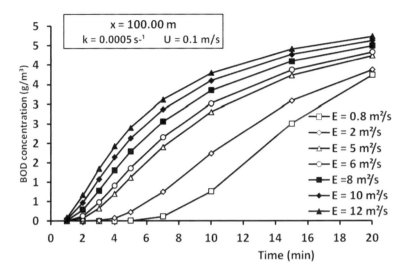

Fig. 20.2 Model output at various values of E

The example considered in the preceding paragraph can help to understand the procedure.

The graph in Fig. 20.2 confirms that the dispersion coefficient E is a parameter to which the model is very sensitive: the pollutant concentration C at the point $x = 100.00$ m, shown as a function of time, remarkably varies after assuming different values of the parameter. Having assumed an initial concentration $C_0 = 7.5$ g/m^3 of BOD, the model gives the concentration at the requested location and time, also expressed in g/m^3. The time interval around $t = 10$ min is clearly that in which the model sensitivity is the highest.

After several runs of the model, the pollutant concentration at $x = 100.00$ m and $t = 10$ min is plotted as a function of E in Fig. 20.3. A difficulty now arises in relation to the calculation of the derivative (20.1), which, being related only to the variable E, is an ordinary derivative and can be written as

$$\underline{C'} = \frac{dC}{dE} \tag{20.2}$$

The difficulty is because these considerations are realised on a numerical procedure, aggravated by the fact that the steps of the variable cannot be considered "little" and are not of constant length. The general theory of numerical calculus (Burden and Douglas-Faires 2000) proposes several tools able to estimate a derivative starting from a set of numerical values or interpreting a graph. In any case, the analytical expression (20.2) is replaced by an expression in finite terms, such as

$$\underline{C'} \cong \frac{\Delta C}{\Delta E} \tag{20.3}$$

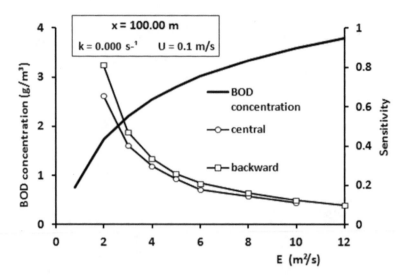

Fig. 20.3 Concentration for different values of E and its derivative

in which ΔE should be sufficiently small.

In the example, as shown in the Figure, the behaviour of the function $C = C(E)$ is sufficiently "smooth", and a numerical procedure is expected to be applied without jumps and discontinuities. As shown in Table 20.3, two of the simplest procedures are applied, namely, the *backward approach*

$$\underline{C'} \cong \frac{C(E_i) - C(E_{i-1})}{E_i - E_{i-1}} \tag{20.4}$$

and the *central approach*

$$\underline{C'} \cong \frac{C(E_{i+1}) - C(E_{i-1})}{E_{i+1} - E_{i-1}} \tag{20.5}$$

Both procedures are comparable, taking into consideration that the sensitivity analysis does not require highly precise results, but aims principally at showing trends and situations worthy of particular attention in the model application.

The content of the table should be read, for instance, in terms of "a unit change in the value of E would change of 0.30 g/m^3 the value of model output".

However, the procedure still looks somewhat difficult to handle, also because the dimension of the derivative leads to a number, whose significance cannot be easily understood. In the case of sensitivity to E, the derivative dimension is [ML^{-5} T], which could give some uncertainty in practice, with the selection of proper units of measurement.

Table 20.3 Model sensitivity to the dispersion coefficient

E (m^2/s)	C (mg/l)	Sensitivity	
		Central approach	Backward approach
0.80	0.76		
2.00	1.74	*0.66*	*0.81*
3.00	2.21	*0.40*	*0.47*
4.00	2.55	*0.30*	*0.34*
5.00	2.81	*0.24*	*0.26*
6.00	3.02	*0.18*	*0.21*
8.00	3.35	*0.14*	*0.16*
10.00	3.60	*0.11*	*0.12*
12.00	3.80		*0.10*

To overcome such a disadvantage, Chapra (1997) and Chapra and Canale (1988) propose the nondimensional term (CN)

$$CN = \frac{p\Delta C}{C\Delta p} \qquad (20.6)$$

called the *condition number* in respect to the parameter p. Obviously, all the terms refer to the particular value p_0 around which the calculations are performed. The condition number can help to improve the procedure of the sensitivity analysis.

The sensitivity evaluation is less precise at the lowest values of time t, during the transition phase of the pollutant transport, and depends on the accuracy characterising the calculations. Shorter intervals of parameter perturbation will give more detailed values. Moreover, it confirms that the transition phase is also that where the effects of a parameter change are more profound.

At the present time, several investigations are in progress, aiming at a better identification and calculation of sensitivity. Statistics provides useful tools, and proper software (ICIM 1992; Bach et al. 1989) is also available to facilitate the practical applications, where sensitivity analysis is included as a special routine of the model development. Comprehensive reviews on the subject are available in Breierova and Choudhari (2001) and Saltelli et al. (2008).

20.3   Uncertainty Analysis

An extension of sensitivity analysis, on which there is now an intensive scientific effort (Radwan et al. 2004; Willems 2010), starts from the consideration that all the model components can be characterised by imprecise values, which can contribute to give an output not completely reliable. In fact, an accurate insight on how the model is developed and works can put into evidence that both the input variables and the parameters can be affected by imprecision and measurement errors, while the model

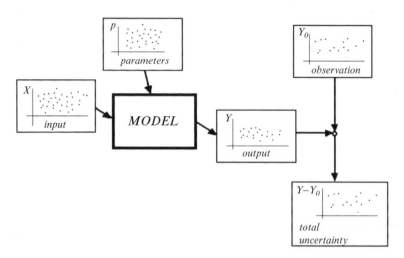

Fig. 20.4 The various sources of uncertainty in a model

itself is based on approximations and mathematical developments that do not interpret correctly the real phenomena to be modelled.

Therefore, a more accurate form of analysis has been developed, *uncertainty analysis*. While sensitivity analysis is based on the assumption of arbitrary values for some model components, uncertainty analysis takes into account that all the components can be characterised by uncertain values and have an impact on the final result.

At a first glance, uncertainty analysis could appear mostly an exercise with poor interest for a practical application in the field. Nevertheless, it can help to test the model capability as a tool able to solve the greatest set of practical problems. This kind of analysis becomes essential if the water quality model is itself the component of a more complex system comprising several models, one another interacting and each one characterised by its own uncertainty. Typical is the interconnection of a water quality model with a hydrological model, which provides the input values in terms of water velocity, or the case where the water quality model provides the basic data for models aiming at assessing a rational use of water resources, as shown in Chap. 17.

The uncertainty analysis works according to the scheme described in Fig. 20.4. The first step is to ascertain the level of inaccuracy for all the terms of the model.

Starting from the input, the main reason of its uncertainty lies in the precision of field measurement. The water velocity is determined by means of the current meter, which gives values depending on its position inside the water body, as well as on some conditions of its operation (internal friction) and its calibration. Similar considerations hold for the pollutant injection, where the concentration of the injected pollutant is determined through laboratory analyses of samples taken in a few points of the river cross section.

Uncertainty analysis starts with the identification of the error ε for the various model components. According to statistics, the error (or *residual*) is now defined as the difference between a measured value x of an entity and its *true* or *theoretically acceptable value* x_0:

$$\varepsilon_i = x_i - x_0 \tag{20.7}$$

As described in Chap. 19, every measured value has its own error.

After calibration and validation and after sensitivity analysis, a first consideration on the model reliability concerns the effect of various uncertainty sources that characterise the terms and the internal structure of the model. In fact, the output model can be affected by errors due to

1. Input variables, which suffer from imprecision and measurement errors (*input error*, ε_{inp})
2. Parameters, which can be chosen improperly (*parameter error*, ε_{par})
3. Model structure (*model structure error*, ε_{mstr})

Moreover, after comparing the model output with direct measures in the field, in the calibration or validation phase, a further error has to be considered, because the model output Y is compared with the field measurement $\cancel{Y}$ that is affected by the proper *observation error* ε_{obs}.

This gives rise to the *final (or total) model error* $\varepsilon_{mod} = (Y - \cancel{Y})$, which should

$$\varepsilon = u = \sqrt{\frac{\sum\limits_{i=1}^{N} (x_i - x_0)^2}{N - 1}} \tag{20.8}$$

be eventually quantified.

Invoking the fundamentals of statistics, each error can be expressed by the relevant standard deviation, whose value can be assumed as the *uncertainty*, u, of the considered entity. In a quite general way, if $x_1, x_2, \ldots, x_N$ is a set of measurements and x_0 a reference value of them, the square of u is the variance of the considered set of x_i and is conventionally identified as $\sigma^2 = u^2$.

In water quality problems, the various errors considered are one another independent, but they contribute to make up the final model error. The theory of component uncertainty says that if ξ_1 and ξ_2 are two terms independent from each other, the variance of their sum is the sum of their variances:

$$\sigma^2(\xi_1 + \xi_2) = \sigma^2(\xi_1) + \sigma^2(\xi_2) \tag{20.9}$$

Applying then these basic concepts to the errors under examination, it follows that

$$\sigma_{mod}^2 = \sigma_{inp}^2 + \sigma_{par}^2 + \sigma_{mstr}^2 + \sigma_{obs}^2 \tag{20.10}$$

being σ_{mod}^2, σ_{inp}^2, σ_{par}^2, σ_{mstr}^2 and σ_{obs}^2, respectively, the variances of errors ε_{mod}, ε_{inp}, ε_{ipar}, ε_{mstr} and ε_{obs} previously identified. Similar considerations were presented in Chap. 19 for data uncertainty.

In line with the above, the model uncertainty is

$$u_{mod} = \sqrt{u_{inp}^2 + u_{par}^2 + u_{mstr}^2 + u_{obs}^2} \qquad (20.11)$$

As mentioned, the errors considered in this analysis refer to the *true* value of the measured or estimated terms, which cannot be achieved and is in practice a meaningless entity. A significant value must be taken that should be reliable and easy to determine.

For a set of generic values x_1, x_2,..., x_N, such a significant value can be the arithmetic mean:

$$x_0 = \frac{\sum\limits_{i=1}^{N} x_i}{N} \qquad (20.12)$$

It depends on the number of measured or estimated values, and it is clear that different sets of these values give different arithmetic means. An accurate examination of this aspect involves the application of the proper tools of statistics, but, for the practical scope of this chapter, it may be acceptable that if the number of measured values is sufficiently great, their relevant arithmetic mean can be stable and, therefore, reliable.

Assuming the arithmetic mean as a reference term, significant values of the errors specified above can be found, and the "goodness" of the model can be defined. Nevertheless, it must be noticed that, while such an assumption is feasible for the input, parameters and observation values, a similar procedure is not applicable for the model structure error, which depends on factors pertaining to mathematical peculiarities that cannot be measured or quantified. This hindrance could be overcome running repeatedly the model with different mathematical approximations, or, in a more general perspective, adopting two or more models based on different procedural assumptions. However, this is indeed costly and time consuming.

It is worthy to mention that the model structure error is the most effective tool to judge the goodness of the model.

According to (20.11), a reasonable approach (Willems 2010; Radwan et al. 2004) suggests that, while all the terms can be measured and the relevant error quantified, u_{mstr}^2 becomes the unknown to be determined.

In the preceding paragraphs, the errors are figures whose significance can be assessed only in accordance with the modeller's experience, who can say whether the error is acceptable or not, depending on his perception of the real aspects of the problem modelled. A way to identify some criteria for a more objective evaluation can be given by an uncertainty analysis in terms of probability, adopting the strict statistics procedures. Referring to the proper textbooks, in the following lines, the basic aspects of such procedures are recalled.

For the generic set of values $x_1, x_2, \ldots, x_N$ already considered, having arithmetic mean x_0 and variance σ^2, the probability that a value x_i falls inside the $[a, b]$ interval, where a and b are prefixed values, is expressed by

$$P(a \leq x_i \leq b) = \int_a^b p(x_i)\mathrm{d}x$$

being $p(x_i)$ the *probability density function* of x_i. The probability P is expressed as percent of the total number of available values and makes up the *cumulative distribution function*.

There are several forms of $p(x_i)$, according to the nature and behaviour of x_i, but for a first approach and in the majority of practical cases, the Gaussian (or normal) probability density function is used:

$$p(x_i) = \frac{1}{\sqrt{2\pi}\sigma} \exp\left(-\frac{(x_0 - \sigma)^2}{2\sigma^2}\right)$$

The above prefixed values a and b can be expressed, respectively, as multiple of the differences $(x_0 - \sigma)$ and $(x_0 + \sigma)$ and give

$$P((x_0 - \sigma) \leq x_i \leq (x_0 + \sigma)) = 68.3\ \%$$
$$P(2(x_0 - \sigma)) \leq x_i \leq 2(x_0 + \sigma)) = 95.5\ \%$$
$$P(3(x_0 - \sigma)) \leq x_i \leq 3(x_0 + \sigma)) = 99.7\ \%$$

These percent values are the *confidence levels* of the x_i measurements.

Applying these considerations to the errors relevant to the model, it is possible to assess their uncertainty and evaluate the reliability of the adopted procedures. In practice, a confidence level of 95% is a good indicator of the model reliability.

20.4   The Monte Carlo Analysis

The methods of statistics can assist in the implementation of the procedures described in the preceding paragraphs. As mentioned, the various model components are characterised by uncertainty due to errors in measurements or estimation, which can contribute to the lack of reliability. It has been repeatedly mentioned that the model runs on significant values of input and parameters, and the model output consists of a value which contains the effect of the errors due to the initial uncertainty and to the mathematical discretisation and interpretation.

If a set of measured values is considered for the input, the significant term to put in the model can be, for instance, their arithmetic mean, which, as mentioned, depends on the considered set. Out of the total number of measurement that can be done for an entity, defined as *population*, there are several *samples* of a limited number of terms. Each sample makes up a set with a proper value of the arithmetic mean. There is, therefore, the need to look for a mean that, as far as possible, takes into account the population of the measurement that can be practically obtained for the considered entity.

Another uncertainty can be due to the choice of parameters. Even though in the majority of cases the parameters are not directly measured, but taken from literature or similar cases, a large variability characterises their value, and their choice can be always determined by a probability level.

As already seen in Chaps. 18 and 19, in the model calibration and validation, the model output is compared with a set of measured values taken in the field, which suffers from the same uncertainty due to the limits characterising their available measurements. It is worthy to remember that, in the majority of cases, the model is a deterministic process, which gives a result that is strictly dependent on the input value and is not tied to any probabilistic consideration.

The intrinsic errors due to the inner model structure can be put into evidence only after changing some mathematical approximation, which is usually an arbitrary choice tied to the modeller's experience. Yet, the model output contains the uncertainties due to its input, which can be altered by the internal approximations. Alteration can give either an increase or a decrease of the input error and may be ascertained after several repeated runs of the model taking significant input values. It is obvious that such a procedure can be cumbersome and costly and can give only poor information on the real causes of errors. Such a disadvantage may be overcome running the model after several changes, modifying, for instance, the step of discretisation of some variables or adopting different integration procedures for the fundamental equation of pollutant transport. This eventually means to change the model.

The deterministic nature of the model suggests applying the procedure illustrated in the preceding pages. The problem is, therefore, moved to the identification of significant values of input, parameters and calibration, pointing out their probability by means of a proper statistical analysis.

Among the tools derived from statistics, the Monte Carlo method (Whitehead and Young 1979) can be very useful as proved by the scientific literature and the numerous practical applications. Several ways of applying the method can be adopted, leading to the *Monte Carlo analysis* and giving rise to complex machinery that can now benefit from the computer facilities. In the following lines, an elementary form of such an analysis is illustrated, leaving details and further developments to the most advanced research in this field (Spear and Hornberger 1983).

Related to the measurable entities used in the model, namely, input, parameters and calibration terms, the Monte Carlo analysis can allow an accurate insight on the peculiarities of the values to be used, focussing on the errors that can occur during the measurement. Statistical evaluations can be performed, not constrained by the limits of the available samples, but closer to the real possibility that the entity can assume, or, in other words, closer to the population of real values.

Table 20.4 Field BOD measures at cross section $x = 0.00$ m

Measurement n.	1	2	3	4	5	6	7	8
BOD (g/m^3)	8.4	9.2	9.5	8.2	7.9	8.8	8.9	9.4

Table 20.5 Development of Monte Carlo analysis for the input data (g/m^3)

Input data (in increasing order)	Probability density function	Cumulative distribution function
7.365	0.038	0.008
7.472	0.057	0.013
7.509	0.065	0.016
7.553	0.076	0.019
7.617	0.094	0.024
7.649	0.104	0.027
7.714	0.127	0.035
7.740	0.138	0.038
7.763	0.147	0.042
7.863	0.193	0.059
7.867	0.195	0.059
8.019	0.280	0.095
8.038	0.292	0.101
8.038	0.292	0.101
8.067	0.310	0.110
...	...	...
...	...	...

The Monte Carlo method considers very large sets of values that a measure can assume, for which the main statistical terms are considered pointing out their probability.

The following example concerns the measure of BOD in a river, for which, in a period of stable hydrological and pollutant injection conditions, the BOD concentration at the cross section $x = 0$ has been measured eight times obtaining the values shown in Table 20.4.

There are several reasons for this discrepancy, among which the difficulty of identifying in the field the time $t = 0$ and locating the collection of the BOD samples in a significant point of the river cross section.

The values are characterised by an arithmetic mean of 8.7875 g/m^3 and a standard deviation of 0.5418 g/m^3.

A simple way of applying the Monte Carlo analysis of the data to be used as model input is illustrated in Table 20.5. Using a spreadsheet, a set of values is randomly generated, taking care that they are selected in accordance with the arithmetic mean and standard deviation of the set of measures, as identified above. To speed up the procedure, the same values are shown in increasing order in the first column of the table. The relevant values of probability density and cumulative distribution are eventually calculated by means of the functions usually available in the spreadsheet and shown, respectively, in column 2 and column 3.

Table 20.6 Statistical terms of the input data (g/m³)

Mean	8.810
Standard deviation	0.605
Max	10.617
Variance	0.3660
Min	7.365
$\mu - \sigma$	8.205
$\mu + \sigma$	9.415
$\mu - 2\sigma$	7.601
$\mu + 2\sigma$	10.020
75 percentile	9.215

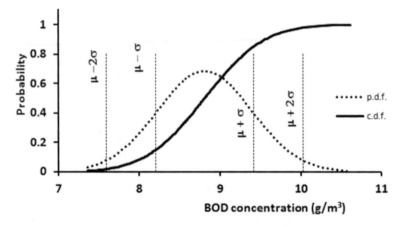

Fig. 20.5 The probability plots of the input data used for the example of Monte Carlo analysis

The procedure can be repeated for a high number of rows, as allowed by the spreadsheet. It is clear that the higher is the number of rows, the closer is the result to the statistical population of the measurement that can be achieved in the field. Usually, some hundreds or thousands of rows are considered, but in the example, for the sake of simplicity and for the explanation purposes, a more small number of rows have been considered satisfactory.

The statistical terms relevant to the values contained in the first column, obtained by means of the functions available in the spreadsheet, are shown in Table 20.6.

As a first remark, it can be noticed that the arithmetic mean and the standard deviation are close, even though not necessarily equal, to respective values inferred from the direct measures in the field and initially assumed for the development of the analysis.

The Monte Carlo analysis provides further information, as it is contained in the same Table 20.6 and shown in Fig. 20.5, in which the curves of probability density (*pdf*) and cumulative distribution (*cdf*) are plotted. Remarkable are also the extreme minimum and maximum that can be expected in the field measurement, as well as the percentiles. In particular, the 75th percentile, that is, 9.215 g/m³, states that,

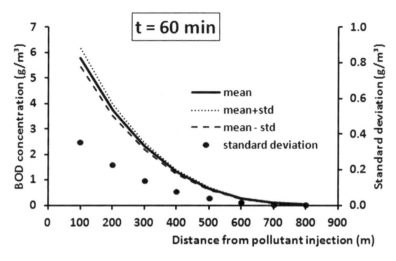

Fig. 20.6 Model output at $t = 60$ min after Monte Carlo analysis

over the total number of possible values, there is a 75% chance of having BOD concentration lower than or equal to 9.215 g/m^3. The intervals $[\mu - \sigma, \mu + \sigma]$ and $[\mu - 2\sigma, \mu + 2\sigma]$, also shown in the Figure, give, respectively, the confidence level of 68.3 and 95.5%, or, in other words, they assert that a measurement between 8.205 and 9.415 g/m^3 can have a 68.3% of probability of occurrence and a measurement between 7.601 and 10.020 g/m^3 can have the 95.5% of occurrence.

The standard deviation is the uncertainty of the input value and gives the measure of error. It can be stated that the BOD measure in the cross section $x = 0$ at $t = 0$ is

$$x_0 = 8.810 \pm 0.605 \text{ g/m}^3$$

The Monte Carlo analysis can also help the calculation of the parameter uncertainty, σ^2_{par}. Nevertheless, since the parameters are usually chosen out of a restricted set of estimates or following similar cases reported in literature, a few values can be satisfactory, to be assumed for repeated runs of the model in order to obtain a significant variance. The modeller's professional experience is essential for the purpose (Fig. 20.6).

Both the figures and the tables put into evidence that the error decreases with the concentration. This allows saying that the uncertainty is greater during the transition stage of the pollutant propagation in the river. Table 20.7, for a selected river cross section, will help to determine the output variance necessary for the calculation of total uncertainty.

The procedure illustrated so far gives the error of output model (or uncertainty), but the phase of model calibration has still to be taken into consideration. As already described in previous pages, both the calibration and validation consist of comparing the model output with some measurements taken in the field at the

Table 20.7 Statistical terms of the model output at $t = 60$ min (g/m^3)

| | $t = 60$ min | | | | | | | |
	$x = 100$ m	$x = 200$ m	$x = 300$ m	$x = 400$ m	$x = 500$ m	$x = 600$ m	$x = 700$ m	$x = 800$ m
Mean	5.7919	3.7480	2.3203	1.3183	0.6565	0.2751	0.0940	0.0257
Standard deviation	0.3533	0.2286	0.1415	0.0804	0.0400	0.0168	0.0057	0.0016
Variance	0.1248	0.0523	0.0200	0.0065	0.0016	0.0003	0.0000	0.0000
Max	6.5955	4.2681	2.6423	1.5012	0.7476	0.3133	0.1071	0.0292
Min	4.5605	2.9512	1.8270	1.0380	0.5169	0.2166	0.0740	0.0202
75 percentile	6.0091	3.8886	2.4073	1.3677	0.6811	0.2854	0.0976	0.0266
$\mu - \sigma$	5.4386	3.5194	2.1788	1.2379	0.6164	0.2583	0.0883	0.0241
$\mu + \sigma$	6.1452	3.9766	2.4618	1.3987	0.6965	0.2919	0.0998	0.0272
$\mu - 2\sigma$	5.0853	3.2908	2.0372	1.1575	0.5764	0.2415	0.0826	0.0225
$\mu + 2\sigma$	6.4984	4.2053	2.6034	1.4791	0.7366	0.3087	0.1055	0.0288

location and time instant corresponding to those simulated by the model. Also these measurements contain uncertainty, which contributes to the final error of the model application. The Monte Carlo analysis can help to find the statistical terms of the measured values to be used for the model calibration and validation. These terms will lead to the calculation of σ^2_{mod}.

The model uncertainty can be, therefore, quantified by means of the variance decomposition procedure already described.

The scientific literature contains description of several cases of successful application of Monte Carlo analysis, carried out starting from the last decades of the twentieth century. The method has been also compared with the described procedure of first-order sensitivity analysis (Burger and Letermaier 1975; Warwick 1997), and small differences were found. Other statistical approaches, as it will be mentioned in the next pages, largely benefit from the Monte Carlo analysis.

20.5 New Trends and Future Developments

Statistical methods can provide other procedures for the model uncertainty analysis, which now attract the researchers' attention. The ultimate goal of these investigations is to find out reliable tools that work easily and give an immediate perception of the model validity. They answer to the primary question of any skilled modeller, who is worried about the capability of his model to correctly simulate the reality.

Particular interest is given now to the tools based on the *statistical inference*, which aim at the definition of the probability of an event A conditioned by the ascertained probability of another event B to which it is tied under certain circumstances. Among these, tools predominates the *Bayesian inference*, which is currently applied in a large set of practical cases (Bolstad 2007) of uncertain events. Several researches are now in progress (Kanso et al. 2003; Beck 1987), with the purpose of transferring the basic concept of the method to the problem of water quality model validation.

The method works on the so called Bayes' theorem, which binds together several forms of probability, relevant to the event A taken into consideration, as well as to the occurrence, or hypothesis, of another event B that can influence the behaviour of A. In a quite general form, it is expressed as

$$P(A/B) = \frac{P(B/A)P(A)}{P(B)} \tag{20.13}$$

In the case of water quality models, the terms of such an expression are to be given the following significance:

$P(A)$, the "prior probability" of the set of values concerning the parameters taken into consideration for the model run

$P(B)$, the "marginal probability" of the set of data collected for the model, regardless the probability of the considered parameters

$P(B|A)$, the "conditional probability" of the set of data that have occurred when the condition imposed by the parameter distribution is considered, to read as the "probability of B under the condition A"

$P(A|B)$, the "posterior probability" of the data, resulting after the distribution of the parameter set is considered

The terms contained in the Bayes' theorem request particular calculation procedures, on which the actual scientific efforts is still searching for a reliable formulation suitable to be applied (Kleidorfer and Rauch 2010). In any case, the posterior probability can be a figure able to quantify how the model can be sensitive to the parameter choice.

Another promising tool based on statistical methods is the Generalised Likelihood Uncertainty Analysis (GLUE), originally developed for hydrological models (Beven and Binley 1992) and recently applied also to water quality models (Freni et al. 2008, 2009; Mannina and Viviani 2009). The basic principle of the procedure is the comparison of the output of several model runs, conceived as different models, which contain proper errors, and fit the observation data. A probabilistic analysis of the results of such comparison leads to the best uncertainty evaluation.

To conclude this short discussion, not necessarily complete, it is worthy to underline that the subject of model uncertainty analysis is the focus of the present and future research. The model structure and its working conditions are by this time entirely developed and familiar in the professional field, as confirmed by the very high number of practical applications. What still requires a better knowledge is the certitude about the model reliability.

References

Bach HK, Brink H, Olesen KW, Havno K (1989) Application of PC-based models in river water quality modeling. Water Quality Institute, Horsholm

Beck MB (1987) Water quality modelling: a review of the analysis of uncertainty. Water Resour Res 23(8):1393–1442

Beven KJ, Binley A (1992) The future of distributed models: model calibration and uncertainty prediction. Hydrol Process 6:279–298

Bolstad WM (2007) Introduction to Bayesian statistics, 2nd edn. Wiley, London/New York

Breierova L, Choudhari M (2001) An introduction to sensitivity analysis, System dynamic in education project. Massachusetts Institute of Technology, Cambridge, MA

Burden RL, Douglas-Faires J (2000) Numerical analysis, 7th edn. Brooks/Cole, Pacific Grove

Burger SJ, Letermaier DP (1975) Probabilistic methods in stream water quality management. Water Resour Bull 11(1):115–130

Chapra SC (1997) Surface water-quality modeling. WCB-McGraw Hill, Boston

Chapra SC, Canale RP (1988) Numerical methods for engineers, 2nd edn. McGraw-Hills, New York

Freni G, Mannina G, Viviani G (2008) Uncertainty in urban stormwater quality modelling: the effect of acceptability threshold in GLUE methodology. Water Res 42:2061–2072

Freni G, Mannina G, Viviani G (2009) Uncertainty assessment of an integrated urban drainage model. J Hydrol 373:392–404

ICIM (1992) A micro-computer package for the simulation of one-dimensional unsteady flow and water quality in open channel system. Bureau ICIM, Rijswijk, p 107

Kanso AM, Gromaire MC, Gaume E, Tassin B, Ghebbo G (2003) Bayesian approach for the calibration of models: application to an urban stormwater pollution model. Water Sci Technol 47(4):77–84

Kledorfer M, Rauch W (2010) Uncertainty analysis in urban drainage modelling. In: Proceedings of the XXXII congress of hydraulics and hydraulic construction, University of Palermo Italy (printed summary)

Mannina G, Viviani G (2009) Parameters uncertainty analysis of water quality models for small river. In: Proceedings of the 18th world IMAC/MODSIM congress, Cairns, Australia, July 13–17, pp 3201–3207

Radwan M, Willems P, Berlamont J (2004) Sensitivity and uncertainty analysis for water quality modelling. J Hydroinform 6(2):83–99

Saltelli A, Ratto M, Andres T, Campolongo F, Cariboni J, Gatelli D, Saisana M, Farantola S (2008) Global sensitivity analysis. Wiley, London

Spear RC, Hornberger GM (1983) Control of DO level in a river under uncertainty. Water Resour Res 19(5)

Warwick JJ (1997) Use of first order sensitivity analysis to optimize successful stream water quality simulation. J Am Water Resour Assoc 33(6):1173–1187

Whitehead P, Young P (1979) Water quality in river systems: Monte Carlo analysis. Water Resour Res 15(2):451–459

Willems P (2010) Model uncertainty analysis by variance decomposition. In: Proceedings of the XXXII congress of hydraulics and hydraulic constructions, Palermo, Italy

Yen BC, Chang ST, Melching CS (1986) First-order reliability analysis. In: Yen BC (ed) Stochastic and risk analysis in hydraulic engineering. Water Resources Publication, Littleton, pp 1–3

Chapter 21
Final Thoughts and Future Trends

Abstract The development of mathematical water quality models is still a matter of intensive research, with the aim of obtaining reliable tools, able to give the closest interpretation of "real world" and suitable to be applied to the wide range of management decisions. Attention is paid primarily to the Fickian models, which are predominant due to their successful applications in a variety of problems. Currently, several attempts are also in progress to develop and test models based on different fundamentals, adopting concepts borrowed from other disciplines. This chapter discusses the future trends in water quality modelling.

21.1 New Developments

As already mentioned in this book, the matter concerning the described water quality models is in continuous evolution. The main interest is to improve the models and make their mechanisms work more and more efficiently, able to give a correct picture of the reality in practical applications and assist experienced professionals in management decisions.

In such a perspective, a lot of efforts have been made for the numerical integration of the fundamental differential equation of pollutant transport. The main aim of these modelling efforts is to produce precise methods easy to handle by means of computing facilities.

The remarkable progress achieved in the use of finite element method is expected to assist in better representation of the complexity of the river geometry. A further step in this direction will be in the improvement achieved by the use of finite volume method, which aims at the development of two- and three-dimensional models. Needless to say that the one-dimensional approach cannot give a satisfactory answer to several cases of large rivers affected by surface plumes or characterised by a stratified pollutant concentration.

In the abundant literature of water quality models, the Fickian approach still remains the firm scientific background for the simulation of pollution transport in

M. Benedini and G. Tsakiris, *Water Quality Modelling for Rivers and Streams*,
Water Science and Technology Library 70, DOI 10.1007/978-94-007-5509-3_21,
© Springer Science+Business Media Dordrecht 2013

rivers and streams. This is so in spite of the criticism that aims at pointing out the faults of the method. In fact, according to some authors, the Fick law does not represent always correctly the behaviour of the pollutant particles in the liquid mass. In particular, some in situ observations confirm that the pollutant in a river does not entirely comply with what an advection-dispersion model can predict (Valentine and Wood 1977; Bottacin-Busolin 2010).

The focus of the more recent investigations is, therefore, brought to the use of in situ observations, highlighting the significance of the terms of the mathematical models.

Some authors (Ganoulis et al. 2001) maintain that an accurate simulation of the water pollution problem is impossible because the data currently available are always incomplete. The measurement of the pollutant concentration in a stream strictly depends on factors that cannot be evaluated, due to the continuous variation in time and space. The dispersion of the velocity field is another reason of imprecision, which cannot be satisfied by the adoption of average values and arithmetic means. This holds both for the variables and the parameters included in the model formulation and leads to the conclusion that the model results should be accepted as indicative values. Under this viewpoint, an attempt to search for an accurate and precise interpretation of the quality problems becomes eventually unfeasible and meaningless.

Such a concern does not vanish the effort for deeper investigations. The attention is now brought to procedures that allow the model user to adopt terms that are describing more closely the complexity of reality.

One of the most interesting attempts to enhance the Fickian approach concerns the use of the *fuzzy sets* (Ganoulis et al. 2003; Shresta et al. 1990; Xiangsan and Chongrong 2001), borrowing a theory now widely applied to the several aspects of the actual life that are characterised by imprecise or insufficient information. The fundamentals of the fuzzy set are described in many textbooks and papers (Klir and Folger 1988). In short, they are based on the assumption that an entity comprises several dispersed values around a more probable figure. This leads to introducing the *fuzzy number*, which is an extension of the regular number (defined therefore as *crisp number*) that does not refer to a single value but to a connected set of possible values. There are numerous papers on the application of fuzzy numbers in water resources problems (Spiliotis and Tsakiris 2007; Tsakiris and Spiliotis 2011; Reddy and Duckstein 1990). According to many authors, the proposed method allows a more realistic evaluation of the uncertainty when the problem is supported by an insufficient set of data. It may be, therefore, considered as an alternative to the uncertainty analysis carried out by means of the statistical procedures described in Chap. 20.

Further, today there are new approaches of formulating the pollutant transport in rivers in a more accurate interpretation of the physical phenomena that govern the pollutant transport. In fact, the pollutant concentration is greatly affected by the irregularities in the river embankments, by the natural bends and by the natural or man-made obstacles that give rise to pockets of stagnant flow, or *dead zones*, in which the pollutant is alternatively entrapped and released. A secondary pollutant transport occurs, defined as *scaling dispersion*, to which the Fickian approach cannot be applied.

The quite general considerations, motivated by in situ observations on the real behaviour of the pollutant particles, suggest that there is a need for a revision of the advection-dispersion approach and the improvement of the Fickian models.

Typical is the case of secondary phenomena occurring in the river, principally in the exchange of pollution between the running water and the sediment at the bottom (Marion and Zaramella 2005; Deng 2002; Cheong et al. 2005). This is the case of the *hyporheic exchange*, for which the Fickian models can give only approximate answers. Nevertheless, some authors maintain that the hyporheic exchange, which involves a scaling dispersion, can still be treated by the Fickian approach (Deng and Jung 2007) after replacing the general equation of pollution transport with a system of two equations, assuming that a pollutant mass of concentration C_R is entrapped in the dead zone for a residence time T_n (Eq. 21.1). The dead zone has a transverse area A_s, while A is the river cross section.

$$\frac{\partial C}{\partial t} + U \frac{\partial C}{\partial x} = K \frac{\partial^2 C}{\partial x^2} + \frac{A_s}{A} \frac{1}{T_n} (C_R - C)$$
$$\frac{\partial C_R}{\partial t} = \frac{1}{T_n} (C - C_R) \tag{21.1}$$

The parameters of these equations have to be determined following experimental measurements.

21.2 Future Trends

More recently, flourishing innovative researches have given rise to the family of *non-Fickian* models, which could better take into account the particle motion, at the same time avoiding the complexities that characterise the application of the advection-dispersion algorithms. These considerations, initially developed for groundwater problems, have now been successfully extended to rivers (Boano et al. 2007). The non-Fickian models deserve a great attention, because they can lead to formulations easily applicable with high level of reliability.

Among these models very promising seem to be those in which the pollutant transport is interpreted by means of the *random walk* theory (Souza 2001), a chapter of statistics that is now currently applied to practical cases characterised by uncertainty. According to the basic principles of this theory, the motion of a pollutant particle is interpreted as a sequence of random jumps with different length and duration, which are treated as random variables with known probability.

There are several ways to explain the random walk principle, according to the particular problem under investigation (Spitzer 2001). According to Fig. 21.1, during the $[t_0 - \tau]$ time interval, a pollutant particle initially located at the point A_0 can randomly jump to several other points in the space around A_0. There is a certain probability that the particle arrives, for instance, at point A_1 at the time $t_0 + \tau$. Such a probability can be expressed as a function of the distance between A_0 and A_1 and the duration τ of the time interval. After developing a proper

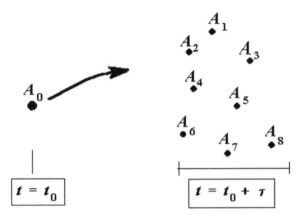

Fig. 21.1 Some possible "jumps" of the pollutant particle

formulation of such probability (Klafter and Silbey 1980), the time evolution of the particle distribution can be estimated as $C(x, t)$, in terms of the quantity of particles that at time t arrive at and depart from location x.

At present, the random walk theory is the subject of systematic research aiming at stressing its validity through practical applications, very often in comparison with the fundamental pollutant transport equation based on the advection and dispersion principle.

Another interesting approach to water quality modelling is based on the *theory of fractals*, borrowed from applications in other disciplines. This theory belongs to the field of mathematics dealing with *chaotic systems*. The fractal is an entity defined as a set of fragmented elements that are part of a whole and are one another interacting through several forms of connections, described by proper algorithms (Mandelbrot 1982; Falconer 2003). The presence of pollutant particles in the river water is then interpreted as a set of such elements. Further, some applications of this type to river water seem to be very promising for simulating the pollutant transport (Kirchner et al. 2000; Moore et al. 1991; Neal et al. 2004; Neal 2004).

In this book, a very concise reference was made on the optimisation models by means of which some river water quality problems can be treated not only in respect to the pollutant transport in the water volume but under a more general perspective of a suitable policy of water resources management and protection (Chi and Harrington 1973). The optimisation procedure is utilised for more complex management problems, where the pollutant transport in the river is a particular problem, usually treated by means of a Fickian model (Maeda and Kawachi 2003). The optimal solution is achieved by conventional optimisation algorithms through an overall objective function that incorporates the minimisation of pollutant concentration. The procedure is now made easier if suitable software packages can be used.

To complete the review on water quality models mention can be done on the more integrated approach to the pollutant transport in river that considers the full process of pollution generation at the river basin scale. This approach does not examine the behaviour of the pollutant in the liquid mass but starts from the pollution generation in relation to identified local sources of contamination and is based on a significant hydrological input, which is responsible for pollutant collection and conveyance along the various pathways and finally in the river. Because the advective-dispersive algorithms are not taken into consideration, the approach can be included within the non-Fickian procedures, giving rise to the *water quality hydrological models* (Rinaldo et al. 2006; Botter et al. 2006; Chuan and Yongming 2001).

The approach is based on a simple balance equation, which refers to the pollutant concentration at the closure cross section of the river basin

$$\frac{\mathrm{d}C}{\mathrm{d}t} = S - A \tag{21.2}$$

in which S is the total source of contamination and A contains the effects of pollution abatement.

The substantial aspect of the problem consists of the proper expression of terms S and A in relation to the geomorphological characteristics of the basin and to the hydrological information considered.

Various tools are used, and the probabilistic analysis is often adopted. The hydrological information gives the water velocity and the propagation time of the flow; chemical and biological interpretations of the pollutant behaviour give its variation along the natural or artificial path of water in the catchment. The model structure and development require complex mathematical formulations, but the most recent applications are able to present answers, related to the pollutant concentration at any cross section of the receiving river. Numerous investigations are in progress in order to improve the current application of these models, which are expected to become promising and reliable in the future.

Beside the model improvement, the refinement of the uncertainty analysis is another subject of research. This is an answer to the common question "How much is the model reliable and what is the confidence level of its output?" The question is always fundamental because, as already repeated in the preceding pages, a lack of confidence on the models still persists among the persons who deal with the water quality problems. Unfortunately, the actual discussion on model reliability is still confined in restricted academic circles, where the available tools are somewhat complex and not easy to handle.

There is, therefore, the need to improve the actual methodologies, pointing out, in particular, at the possibility to put them in the hands of practical operators who are not necessarily expert on statistics and computational mathematics but have a clear view of the water quality behaviour in the water body for the management and protection of which they are responsible.

The improvement of the model structure and the refinement of the sensitivity analysis are parallel with the search for more systematic and reliable data, and the researcher's interest is oriented also towards the procedures of data collection. As already shown at the beginning of these pages, running a model and refining its output are also efficient ways to learn what data are the most appropriate and useful to solve the problem. Searching for new data means also improving the equipment for in situ measurements and defining how the measurements have to be carried out in time and space.

An important point of data collection concerns the development of monitoring activities, which allow several types of data, of different nature but related to the problem under examination, to be simultaneously collected and processed. Advanced projects have been also completed (Strobl and Robillard 2006), involving methods and tools of the artificial intelligence, namely, the *expert systems* (Fedra 1993), the *artificial neural networks*, the *genetic algorithms* (Harrell and Ranjithan 1997) and the already mentioned fuzzy logic systems (Tsoukalas and Uhrig 1997). Monitoring is, therefore, a complementary activity of the model application.

A review of the state of the art should also mention that today there is a flourishing market of software packages, in which the model algorithms are embedded, very often in a sealed form and without sufficient explanation. The model is therefore presented as a black box, to be used with the proper input data and able to give the desired output, also in situations characterised by a great number of variables. Since the operator is not requested to consider the internal structure of the model, the model itself can be put in the hands of the largest set of people, which could become acquainted with this powerful tool.

The integrated approach based on the entire river basin is promoted strongly by the Water Framework Directive of the European Union. It is expected that during the implementation of WFD, systematic collection of water quality data will take place in all European countries. New water quality models based mostly on the entire river basin will be devised to fulfil the obligations resulted from the WFD implementation.

In this perspective, integration obtained by geographical information systems (GIS) is expected to play an important role for characterising the spatial variability of key quality variables in each river basin (WFD 2002). Finally, advances in remote sensing technology are expected to assist in water quality data collection, mainly in the areas where the difficulty of collecting infield data has given insufficient information.

21.3   Epilogue

As a final statement, the water quality problems in river and streams remain one of the most fundamental problems in water resources management. The mathematical models are the most effective methods to understand the behaviour of pollutant transport.

Regarding the uncertainty of the results produced by existing models, a systematic attempt is currently undertaken for devising new models and procedures that aim at lowering the uncertainty and increasing the engineering confidence.

The success of mathematical models relies on the way the real and human-induced processes are understood and formally interpreted. This can be achieved only with a skilled capability, essential not only to develop the model but also to assess the results of an application and put them into management decisions. In such a way, the mathematical models for river water quality can give a very important contribution to the enhancement of the expertise that is necessary for the operational water resources management.

References

Boano F, Packman AI, Cortis A, Ravelli R, Ridolfi L (2007) A continuous time random walk approach to the stream transport of solutes. Water Resour Res 43:w10425

Bottacin-Busolin A (2010) Transport of solutes in streams with transient storage and hyporheic exchange. Ph.D. thesis, University of Padua, Padua, Italy

Botter G, Settin T, Marani M, Rinaldo A (2006) A stochastic model for nitrate transport and cycling at basin scale. Water Resour Res 42:W04415

Cheong TS, Younis BA, Seo IW (2005) Effect of sinuosity on hyporheic exchange in natural streams of rivers. In: Proceedings of the 31st IAHR congress, vol 1. Seoul, Korea, p 335 (printed summary)

Chi T, Harrington JJ (1973) A water quality management model for the Tiber Basin. Italian Water Research Institute, Technical report of META SYSTEM INC., Cambridge, MA, USA

Chuan L, Yongming S (2001) A method for estimating the total organic carbon transport in world rivers. In: Proceedings of the 29th IAHR congress, Beijing, China, theme B, pp 122–130

Deng ZQ (2002) Theoretical investigation into dispersion in natural rivers. Dissertation at the Lund University, Lund, Sweden

Deng ZQ, Jung HS (2007) Scale-dependent dispersion in rivers. In: Proceedings of the 32nd IAHR congress, vol 2. Venezia, Italy, p 571 (printed summary)

Falconer K (2003) Fractal geometry: mathematical foundations and application. Wiley, Bognor Regis, UK

Fedra K (1993) Expert systems in water resources simulation and optimisation. In: Marco JB et al (eds) Stochastic hydrology and its use in water resources simulation and optimisation. Kluwer Academic Press, Dordrecht

Ganoulis J, Anagostopoulos P, Mpimpas H (2001) Fuzzy logic-based risk analysis of water pollution. In: Proceedings of the 29th IAHR congress, Beijing, China, theme B, pp 169–175

Ganoulis J, Anagostopoulos P, Mpimpas H (2003) Fuzzy numerical simulation of water quality. In: Proceedings of the 30th IAHR congress, Thessaloniki, Greece, theme B, pp 165–174

Harrell J, Ranjithan S (1997) Generating efficient watershed management strategies using genetic algorithm-based method. In: Proceedings of the 24th annual conference on water resources planning and management. ASCE, Houston, TX, USA

Kirchner JW, Feng X, Neal C (2000) Fractal stream chemistry and its implications for contaminant transport in catchments. Nature 403:524–527

Klafter J, Silbey R (1980) Derivation of the continuous-time random walk equation. Phys Rev Lett 44(2):55–58

Klir G, Folger T (1988) Fuzzy sets, uncertainty and information. Prentice Hall, Upper Saddle River

Maeda S, Kawachi T (2003) Integrated management of river water quality under uncertainty using optimization model. In: Proceedings of the 30th IAHR congress, Thessaloniki, Greece, theme B, pp 175–182

Mandelbrot BB (1982) The fractal geometry of nature. W.H. Freeman and Company, San Francisco, US

Marion A, Zaramella M (2005) Diffusive behaviour of bedform-induced hyporheic exchange in rivers. J Environ Eng 131(9):1260–1266

Moore ID, Grayson RB, Ladson AR (1991) Digital terrain modelling: a review of hydrological, geomorphological and biological applications. Hydrol Process 5:3–30

Neal C (2004) The water quality functioning of the upper River Severn, Plynlimon, mid-Wales: issues of monitoring, process understanding and forestry. Hydrol Earth Syst Sci 8(3):521–532

Neal C, Neal M, Hill L, Wicham H (2004) The water quality of River Thame in the Thames Basin of south/south-eastern England. Sci Total Environ 360:254–271

Reddy KR, Duckstein L (1990) Fuzzy reliability in hydraulics. In: Proceedings of the 1st international symposium on uncertainty models and analysis, University of Maryland, College Park, MD, USA

Rinaldo A, Botter G, Bertuzzo E, Ucceli A, Settin T, Marani M (2006) Transport at basin scale: theoretical framework. Hydrol Earth Syst Sci 10:19–29

Shresta BP, Reddy KR, Duckstein L (1990) Fuzzy reliability in hydraulics. In: Proceedings of first international symposium on uncertainty models and analysis. University of Maryland, College Park Maryland

Souza R (2001) Simulation of dispersion of pollutant by eddy field velocity. In: Proceedings of the 29th IAHR congress, Beijing, China, theme B, pp 94–99

Spiliotis M, Tsakiris G (2007) Minimum cost irrigation network design using interactive fuzzy integer programming. J Irrig Drain Eng (ASCE) 133:242–248

Spitzer F (2001) Principles of random walk, 2nd edn. Springer, Vienna

Strobl RO, Robillard PD (2006) Artificial intelligence technologies in surface water quality monitoring. Water Int 31(2):198–209

Tsakiris G, Spiliotis M (2011) Planning against long term water scarcity: a fuzzy multicriteria approach. Water Resour Manage 25(4):1103–1129

Tsoukalas LK, Uhrig RE (1997) Fuzzy and neural approaches in engineering. Wiley, New York

Valentine EM, Wood IR (1977) Longitudinal dispersion with dead zones. J Hydraul Div 103 (9):975–990

WFD CIS Guidance Document No. 9 (2002) Implementing the Geographical Information System Elements (GIS) of the water framework directive. Published by the Directorate General Environment of the European Commission, Brussels, ISBN No. 92-894-5129-7

Xiangsan W, Chongrong L (2001) Forecast model of pollutional load in Xiangxi River basin and its application. In: Proceedings of the 29th IAHR congress, Beijing, China, theme B, pp 232–236

Appendix

The following selected water quality models are briefly described: MONERIS, MIKE-11, QUAL2K, SIMCAT, TOMCAT, TOPCAT, IWA RWQM and DRAINMOD. It should be stressed that some of them are not strictly limited to in-stream water quality modelling but also simulate water quality in the corresponding watershed. They are included to show the tendency for more integrated approach towards water quality, which is also promoted by the Water Framework Directive (WFD) of the EU.

MONERIS

The MONERIS (MOdelling Nutrient Emissions in RIver Systems) model builds over eight sub-models that simulate the main processes of generation and transport of suspended solids and nutrients to the river network (Behrendt et al. 2007; Palmeri et al. 2005). Since many countries are moving towards a watershed-based approach as proposed by the WFD, the MONERIS model is configured to support environmental studies in a watershed context (Venohr et al. 2009). MONERIS is a GIS-oriented model which is based on the empirical approach to describe complex relationships in a simple way. It is a conceptual model for the quantification of nutrient emissions from point and non-point sources in river catchments larger than 50 km^2.

The model includes a scenario manager in order to calculate the effects of measures on the emissions of nutrients for different spatial units and pathways.

The main processes and pathways in MONERIS model are shown in Fig. 1: (a) groundwater, (b) erosion, (c) overland flow (dissolved nutrients), (d) tile drainages, (e) atmospheric deposition on water surface, (f) urban areas and (g) point sources (e.g. waste water treatment plants) (Behrendt et al. 2000).

MONERIS seems to be close to become an acceptable tool for solving water quality problems required for the implementation of the Water Framework Directive (European Commission 2002).

M. Benedini and G. Tsakiris, *Water Quality Modelling for Rivers and Streams*, Water Science and Technology Library 70, DOI 10.1007/978-94-007-5509-3, © Springer Science+Business Media Dordrecht 2013

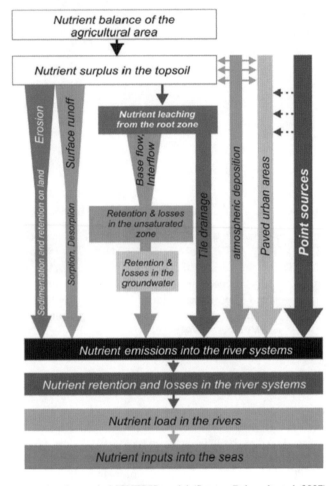

Fig. 1 Processes and pathways in MONERIS model (Source: Behrendt et al. 2007)

MIKE-11

MIKE-11 is a one-dimensional hydrodynamic model that simulates the dynamic water movement in a river (DHI 1998). The model is well suited to complex systems and has been applied as a water quality model to rivers in northern India and England (Crabtree et al. 1996; Kazmi and Hansen 1997). It requires large amounts of data and remains a deterministic model. It consists of a set of modules including sediment transport, water quality, advection-dispersion, rainfall-run-off and eutrophication. It can be used both as a water quality model to assess the impact of intermittent discharges on rivers and as a hydraulic model by flood defence analysts and designers.

The hydrodynamic module, which is the core of the MIKE-11 model, simulates dynamic flows in rivers. This module assumes that the flow conditions are homogeneous within the channel. According to DHI (1998), it solves the full hydrodynamic equations of Saint-Venant (Eqs. 1 and 2):

$$\frac{\partial A_x}{\partial t} + \frac{\partial Q}{\partial x} = q \tag{1}$$

$$\frac{\partial Q}{\partial t} = gA(S_0 - S_f) - gA\frac{\partial y}{\partial x} - \frac{\partial\left(\frac{aQ^2}{A}\right)}{\partial x} \tag{2}$$

in which Q is the flow, t is the time, a is the momentum coefficient, A is the wetted area (or reach volume per unit length), y is the depth, x is the distance downstream, g is the acceleration due to gravity, q is the lateral inflow, S_0 is the bed slope and S_f is the friction slope.

The advection-dispersion module is suitable for the simulation of first-order decays of pollutants. The water quality module provides advanced simulation at one of six different levels of complexity that can be summarised as follows:

(a) First level: first-order decay of dissolved oxygen and BOD plus temperature effects and reaeration
(b) Second level: as the first level plus the exchange reactions between BOD and the sediments
(c) Third level: as the second level plus nitrate and ammonium balances without denitrification
(d) Fourth level: as the third level plus denitrification
(e) Fifth level: as the forth level plus an oxygen demand due to the BOD that settles, but without the nitrogen components of nitrate and ammonium
(f) Sixth level: as the fifth level plus nitrate and ammonium

The processes of coliforms and phosphorus may also be added to any of the six levels. The water quality module can be run with or without sediment processes. It includes heavy metals, iron oxidation, eutrophication and nutrient transport. The advection-dispersion, hydrodynamic and water quality parameters are entered into the system by using a series of editors in the interface. Abstractions, discharges and tributaries are referred to as boundary conditions and are linked to a time series of quality and flow entered in an interface editor.

A water quality file includes the water quality parameters and the default values. The list of coefficient and water quality parameters include, among others, degradation of BOD, influence of dissolved oxygen concentration on the BOD decay, settling velocity of BOD, resuspension of BOD from the bed, oxygen demand by nitrification per unit mass of ammonium converted to nitrate, rate of ammonium release, nitrification rate constant, concentration dependence of nitrification and maximum heat radiation from the river.

Before the simulation of water quality and solute transport processes, MIKE-11 completes the flow simulation. After the successful completion of the flow simulation, the model can be run for water quality using existing hydrodynamic results. The calculations of the mass balance for the water quality indicators are performed at all time steps for all reaches, using the advection-dispersion module.

The MIKE-11 output includes time series of depth, flow and pollutant concentration for in any reach, as well as statistical options to describe the results.

QUAL2K

QUAL2K is a one-dimensional, steady-state model of water quality and in-stream flow. It is neither stochastic nor dynamic simulation model. The QUAL2K model can simulate up to 16 water quality indicators along a river and its tributaries (Brown and Barnwell 1987). The river reach is divided into a number of subreaches of equal length. The model uses the following assumptions:

- The advective transport is based on the mean flow.
- The water quality indicators are completely mixed over the cross section.
- The dispersive transport is correlated with the concentration gradient.

The model allows the user to simulate any combination of the following indicators:

(a) Dissolved oxygen
(b) Temperature
(c) Phosphorus
(d) Nitrate, nitrite, ammonium and organic nitrogen
(e) Chlorophyll-a
(f) Up to three conservative solutes
(g) One nonconservative constituent solute
(h) Coliform bacteria

Furthermore, nitrate, phosphate and dissolved oxygen are represented in more detail, while the majority of indicators are simulated as first-order decay.

The QUAL2K model is suitable for modelling pollutants in freshwater that rely on sediment interactions, especially as a sink of inorganic and organic substances.

The initial step of the standard QUAL2K is to divide the river system into reaches (up to 50), and each of these is then divided into a number of computational elements of equal length. The data requirements of the model in terms of flow and water quality include single values of each pollutant modelled.

The model allows up to 20 computational elements per reach. The main feature of this model is the river reach. The data required for each river reach include

(a) Flow and hydraulic terms
(b) Initial conditions

(c) Reaction rate coefficients
(d) Local climatological values for heat balance computations
(e) Rate parameters for all of the biological and chemical reactions

The output of the model includes the pollutant mass balance and the flow for each reach. The main advantages of QUAL2K are the availability of simulation of algae (chlorophyll-a), an extensive documentation of its code and theoretical background and the availability of free download from its website. The model requires a small amount of data to represent the sediments and only partial hydraulic terms.

SIMCAT

SIMCAT (SIMulation of CATchments) is a one-dimensional, deterministic, steady-state (i.e. time-invariant) model that represents the fate and transport of solutes in the river and the input from point-source effluent discharges on the basis of the following types of behaviour: (a) dissolved oxygen is represented by a relationship involving temperature, reaeration and BOD decay; (b) nonconservative substances which decay (i.e. nitrate and BOD); and (c) conservative substances which are assumed not to decay.

The application of the SIMCAT model requires the splitting of the river system into user-defined reaches, which are generally identified as the distance between points of interest, or tributary junctions (Warn 2010).

Up to 600 reaches can be modelled, and up to 1,400 "features", such as discharges and abstractions, can be included. For each run, the model randomly selects values of quality and flow from the given distributions for all the inputs (Warn 2010).

The outputs of the SIMCAT are summary statistics (90th or 95th percentiles and means) for each reach and each pollutant.

The model has been successfully used in the United Kingdom and is a satisfactory water quality management tool, even though it does not rely on sediment interactions. Another limitation of this model is that it cannot represent temporal variability and is not particularly suitable for complex scenarios. Nonetheless, SIMCAT is a simple versatile model which can be used for reconnaissance analysis of water quality in a river system.

TOMCAT

The TOMCAT model (Temporal/Overall Model for CATchments) makes use of Monte Carlo analysis and was developed to assist in the process of reviewing effluent quality standards at sampling sites in order to meet the objectives of surface water quality preservation (Bowden and Brown 1984).

The model allows for more complex temporal correlations, taking into account the seasonal and diurnal effects in the flow data and the recorded water quality and then reproduces these effects in the simulated data. A number of pollution events that are specified at the rivers, monitoring sites, abstractions and effluent discharges define the river system in TOMCAT.

These events are associated with the following processes: (a) mass balance and flow mixing, (b) input, and (c) internal transformations and additions to flow from groundwater and run-off. Regarding the processes, the flow model in TOMCAT uses the same load additions and simple flow used by SIMCAT.

The types of input data which are required to run the model include (a) fixed values, generally physical parameters that define the reaction rates, and (b) quality data and the flow that are used by the model as input to the process equations.

The TOMCAT structure is defined by

- "Seasons" to be included in the model
- Model runs to be carried out
- Sub-catchments to be simulated and mean monthly air temperatures

According to Cox (2003), a number of parameters are supplied for each user-defined reach including (a) concentration and decay rate of both BOD and NH_4^+, (b) oxygen exchange rate, (c) scale factor for run-off, (d) thermal equilibrium rate constant, (e) catchment number for estimating the diffuse catchment run-off, (f) scale factor for run-off, (g) reach length and depth and (h) mean cross-sectional area.

For the model calibration, the observed data are also usually included, while the values are randomly selected by the model from the flow and quality statistical distributions.

The model is able to simulate the action of storm water overflows. TOMCAT calculates the quality and flow in each reach by solving the process equations.

The recorded data from monitoring stations are compared with simulated flows in the river. Moreover, the simulated flows are automatically calibrated against the recorded data.

The model can be easily applied for simulating the current conditions of flow and water quality in the catchment (e.g. the TOMCAT can be used to predict what changes to the discharge would be required in order to meet legislative criteria in the river). It is designed to be easily set up, while the produced output is suitable for comparison with the values included in the corresponding legislation.

TOPCAT-NP

TOPCAT model uses subsurface flow equations and identical soil moisture stores; it is a minimum information requirement (MIR) version of TOPMODEL (Quinn and Beven 1993). A nitrate model (N-MIR) and phosphorus model (P-MIR) constitute TOPCAT-NP model, which is able to predict hourly or daily nitrate

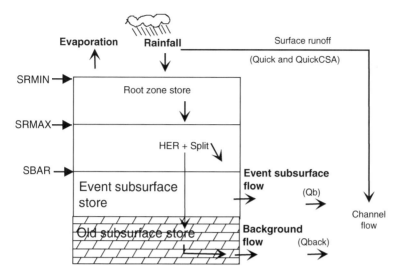

Fig. 2 Diagram presenting the parameters and flow components used in the TOPCAT model [quick is fraction of rainfall in one time step that converts directly into quick surface run-off (0.05 < Quick < 0.3); quick CSA is critical source area quick flow; Q_b is event subsurface flow; Q_{back} is background flow] (Redrawn from Quinn et al. 2008)

and phosphorus losses (Quinn et al. 1999). This model is applicable at a research scale within instrumented sites (1–10 km^2) and more generally at the larger catchment scale (100–1,500 km^2).

The TOPCAT-NP model always tries to retain some physical meaning for the parameters used. TOPCAT and TOPMODEL use the same subsurface flow equations and soil moisture stores but do not use a topographic distribution function (Beven et al. 1995; Quinn et al. 2008). This model contains two overland flow components that are caused by extensive agricultural management practices and an extra baseflow weather flow component. It uses a simple moisture root zone store to receive inputs of potential evapotranspiration and recorded rainfall (Quinn et al. 2008). The moisture content of the root zone store can range from SRMIN to SRMAX (Fig. 2).

Another important parameter of the TOPCAT model is the hydrologically effective rainfall (HER) representing the excess rainfall which is allowed to percolate deeper into the soil. According to Quinn et al. (2008), typically the value of observed root zone depth is more than the value of SRMAX, while SRMAX is an effective "antecedent drying" parameter that triggers the release of HER.

The moisture stores that are used in the TOPCAT model are as follows:

- The unsaturated root zone store
- The saturated "event" subsurface store
- The old subsurface store (or the background flow store)

Finally, a fraction of water of hydrologically effective rainfall (HER) could bypass the event subsurface store and enter the old subsurface store (Fig. 2). The proportion of the HER that enters the event subsurface store is controlled by the parameter SPLIT. SBAR describes the current moisture status in the event subsurface store.

The following formula expresses the rate at which moisture is lost from the store per unit time:

$$Q_b(t) = \exp^{-\gamma} \exp^{SBAR(t)/m} \qquad (5)$$

in which Q_b is event subsurface flow, γ is the mean of the soils/topographic index and is calculated according to the expression $\gamma = \lambda - \ln(T_0)$, λ is the mean of the TOPMODEL dimensionless topographic index, T_0 is the average transmissivity of the subsurface event store when the soil is just saturated and m is the recession rate parameter.

The total amount of water in the subsurface store within each time step is determined by calculating the vertical flow entering the subsurface store (HER × SPLIT) and the amount leaving Q_b

$$SBAR(t) = SBAR(t - 1) - (HER \times SPLIT) + Q_b(t) \qquad (6)$$

where $SBAR(t - 1)$ and $SBAR(t)$ are the catchment storage deficits at the previous and current time steps, respectively.

The original paper published by Quinn et al. (2008) presents a series of equations for the calculation of T-N and T-P concentration in the stream. The TOPCAT-NP has been applied in Glen catchment-Denmark, in the Swan-canning Estuary-Australia and in Bollington Hall-UK (Quinn 2004; Quinn et al. 2008).

DRAINMOD

The software code of DRAINMOD is developed to assist in the simulation of the transport of water and transport and transformation of nitrogen in a stream. DRAINMOD was first used as a research tool to investigate the performance of subirrigation and drainage systems and their effects on contaminant treatment, crop response and water use. The most recent DRAINMOD PC version is the 6.1 version released on 26 October 2011. DRAINMOD has been extended to predict the movement of nitrogen (DRAINMOD-N) and salt (DRAINMOD-S) in shallow water table soils (Deal et al. 1986; Massey et al. 1983).

It should be noted that the effects of frozen conditions on soil water processes as well as snow and snowmelt are not considered in DRAINMOD. Therefore, application is confined to periods when the soil is not frozen, while the model is designed for use in humid regions. Moreover, the DRAINMOD model may be applied for irrigated arid regions where drainage is required and the water table is shallow (Skaggs 1981; Skaggs et al. 1981).

According to El-Sadek (2007), DRAINMOD model allows calculating at the scale of an individual field, the daily nitrate leaching for a given crop, farming, soil and geohydrological conditions. The required information for the application of the model include (a) soil type, (b) water and nitrogen status of the soil profile at the start of the period to be simulated, (c) land use, (d) climate and (e) nitrogen fertiliser practice. Versions of DRAINMOD model have been successfully applied to assess the nitrate-nitrogen leaching in the catchment basin of the river Dender which is located in the west of Brussels. The value of the nitrate load is based on a given land management practice and can be obtained from simulations using DRAINMOD-N for a field of interest. The attenuation rate and the values of nutrient decay may vary with both the location in the river and the season. Generally, lower values are to be considered during high flows, and vice versa (El-Sadek 2007).

DRAINMOD-N and DRAINMOD-S models use a modified version of DRAINMOD to determine average daily soil water contents and water fluxes. The transport of pollutant is determined by an explicit solution to the advective-dispersive-reactive equation.

DRAINMOD-N is able to predict nitrogen concentrations in the soil profile and in surface and subsurface drainage by using functional relationships to quantify rainfall deposition run-off, losses, drainage losses, denitrification, plant uptake, net mineralisation and fertiliser dissolution.

IWA River Water Quality Model

The River Water Quality Model (RWQM) has been developed and presented by the Task Group on River Water Quality of the International Water Association after a critical evaluation of the current state of the practice in water quality modelling. In fact, the majority of the existing models rely on the assumption that BOD is the primary state variable; despite it does not include all the biodegradable matter. This involves a poor representation of benthic flux terms, and, consequently, it is impossible to close completely the mass balances. This limitation in current river water quality models impairs their predictive ability in case some basic river characteristics, such as the pollutant load, the streamflow and the morphometry, undergo a remarkable change. The RWQM is intended to overcome the deficiencies in traditional water quality models, particularly to close the mass balance between the water column and sediment (Borchardt and Reichert 2000; Shanahan et al. 2000). The model is compatible and can be linked with several existing models developed by IWA. Several fundamental water quality components and processes are dealt with by the model, like the cycles of carbon, oxygen, nitrogen and phosphorus, besides the biochemical oxygen demand (Reichert et al. 2000).

RWQM model is developed following the usual assumption of pollutant transport. Particular attention is given to the sediment processes, which are considered not only along the vertical column but also in lateral directions. The algorithm includes the reaction of the main pollution indicators, both in a dynamic situation and in a steady state. An application of uncertainty analysis completes the procedure (Reichert 1995).

A detailed operation manual is available in order to facilitate the practical application (Reichert et al. 2001).

References

Behrendt H, Huber P, Kornmilch M, Opitz D, Schmoll O, Scholz G, Uebe R (2000) Nutrient emissions into river basins of Germany. UBA-FB, 99-087/e

Behrendt H, Venohr M, Hirt U, Hofmann J, Opitz D, Gericke A (2007) The model system MONERIS version 2.0 user's manual. Leibniz Institute of Freshwater Ecology and Inland Fisheries in the Forschungsverbund Berlin e.V., Müggelseedamm 310, D-12587 Berlin, Germany

Beven K, Lamb R, Quinn P, Romanowicz R, Freer J (1995) TOPMODEL. In: Singh VP (ed) Computer models of watershed hydrology. Water Resources Publications, Highlands Ranch, pp 627–668

Borchardt D, Reichert P (2000) River water quality model no. 1: compartmentalisation approach applied to oxygen balances in the river Lahn (Germany). Water Sci Technol 43(5):41–49

Bowden K, Brown SR (1984) Relating effluent control parameters to river quality objectives using a generalised catchment simulation model. Water Sci Technol 16:197–205

Brown LC, Barnwell TO (1987) The enhanced stream water quality models QUAL2E and QUAL2E-UNCAS: documentation and user manual. Athens, GA: US EPA; EPAy 600y3-87y007

Cox BA (2003) A review of currently available in-stream water quality models and their applicability for simulating dissolved oxygen in lowland rivers. Sci Total Environ 314–316:335–377

Crabtree B, Earp W, Whalley P (1996) A demonstration of the benefits of integrated wastewater planning for controlling transient pollution. Water Sci Technol 33(2):209–218

Deal SC, Gilliam JW, Skaggs RW, Konyha KD (1986) Prediction of nitrogen and phosphorus losses as related to agricultural drainage system design. Agric Ecosyst Environ 18:37–51

DHI (1998) MIKE 11: a microcomputer based modeling system for rivers and channels. Reference manual of the Danish Hydraulic Institute, Hoersholm, Denmark

El-Sadek A (2007) Upscaling field scale hydrology and water quality modelling to catchment scale. Water Resour Manage 21:149–169

European Commission (2002) Common implementation strategy for the water framework directive. Guidance document no. 3, analysis of pressures and impacts. The Directorate General Environment of the European Commission, Brussels

Kazmi AA, Hansen IS (1997) Numerical models in water quality management: a case study for the Yamuna river (India). Water Sci Technol 36(5):193–200

Massey FC, Skaggs RW, Sneed RE (1983) Energy and water requirements for subirrigation vs. sprinkler irrigation. Trans ASAE 26(1):126–133

Palmeri L, Bendoricchio G, Artioli Y (2005) Modelling nutrient emissions from river systems and loads to the coastal zone: Po river case study, Italy. Ecol Model 184:37–53

Quinn P (2004) Scale appropriate modelling: representing cause and effect relationships in nitrate pollution at the catchment scale for the purpose of catchment scale planning. J Hydrol 291:197–217

Quinn PF, Beven KJ (1993) Spatial and temporal predictions of soil moisture dynamics, runoff, variable source areas and evapotranspiration for Plynlimon, mid-Wales. Hydrol Process 7:425–448

Quinn PF, Anthony S, Lord E (1999) Basin scale nitrate modelling using a minimum information requirement approach. In: Trudgill S, Walling D, Webb B (eds) Water quality: processes and policy. Wiley, Chichester, pp 101–117

Quinn PF, Hewett CJM, Dayawansa NDK (2008) TOPCAT-NP: a minimum information requirement model for simulation of flow and nutrient transport from agricultural systems. Hydrol Process 22:2565–2580. doi:10.1002/hyp. 6855

Reichert P (1995) Design techniques of a computer program for the identification of processes and the simulation of water quality in aquatic systems. Environ Softw 10(3):199–210

Reichert P, Borchardt D, Henze M, Rauch W, Shanahan P, Somlyódy L, Vanrolleghem P (2000) River water quality model no. 1: II biochemical process equations. Water Sci Technol 43(5):11–30

Reichert P, Borchardt D, Henze M, Rauch W, Shanahan P, Somlyody L, Vanrolleghem A (2001) River water quality model no.1, scientific and technical report no. 12, IWA Publication. London UK, ISBN: 9781900222822

Shanahan P, Borchardt D, Henze M, Rauch W, Reichert P, Somlyódy L, Vanrolleghem P (2000) River water quality model no. 1: modelling approach. Water Sci Technol 43(5):1–9

Skaggs RW (1981) DRAINMOD reference report. Methods for design and evaluation of drainage-water management systems for soils with high water tables. USDA-SCS, 329 pp

Skaggs RW, Fausey NR, Nolte BH (1981) Water management evaluation for North Central Ohio. Trans. of American Society of Association Executives, Washington DC US, 24(4):922–928

Venohr M, Hirt U, Hofmann J, Opitz D, Gericke A, Wetzig A, Ortelbach K, Natho S, Neumann F, Hurdler J (2009) The model system MONERIS version 2.14.1vba manual. Leibniz Institute of Freshwater Ecology and Inland Fisheries in the Forschungsverbund Berlin e.V., Müggelseedamm 310, D-12587 Berlin, Germany. Available at: www.icpdr.org

Warn T (2010) SIMCAT 11.5 a guide and reference for users. Environment Agency. www.environment.agency.gov.uk/static/documents/. . ./111_07_SD06.pdf

Index

A

Accuracy order, 130, 145
Active microzone, 232
Adduction coefficient, 204
Advection (convection), 29–30, 35, 37, 40,
    49, 60, 95, 96, 110, 112, 116, 117,
    127, 128, 130–135, 150, 177, 187,
    200, 201, 204, 205, 210, 266–268
Algae, 70–73, 76–81
Ammonia
    nitrogen, 76–77
    oxidation, 74
Anaerobic condition, 106
Analogical model, 12
Analytical solution, 44, 91–92, 97, 147, 171,
    185, 187

B

Bacteria, 13, 14, 27, 32, 39, 57, 59, 61, 70, 72,
    75, 76, 80, 82, 122, 240
Basic algorithm, 150–160
Bayes' theorem, 261, 262
Benthos, 76, 77, 79
Biochemical oxygen demand (BOD), 13, 39,
    40, 58–62, 64, 69–71, 73, 105, 106,
    116–118, 120–122, 233, 249, 257, 259
    bottle, 61
    decay, 106
    ultimate, 61, 62
Biofilm, 233
BOD. *See* Biochemical oxygen demand (BOD)

C

Catchment area, 23
Chance constrained model, 219

Chaotic system, 268
Chemical oxygen demand (COD), 13, 73, 233
Chlorophyll, 72
Coliform, 80
Computing procedures, 98–100, 218
Concentration, 15, 28, 37, 50, 60, 69, 92, 103,
    110, 127, 150, 180, 186, 200, 213,
    232, 246, 265
    gradient, 31
Conductivity, 8, 13
Confidence level, 235, 236, 238, 239, 255,
    259, 269
Conservation law, 42, 179, 180
Conservative (persistent) pollutant, 28, 38,
    94, 99, 131, 146, 169, 172, 173, 180
Constants, 19, 31, 38, 46, 50, 51, 62, 80–81,
    91–93, 95, 97, 103, 104, 110, 115, 117,
    122, 131, 133, 135, 138, 145, 148, 151,
    189, 192, 194, 195, 209, 215, 217, 225,
    246, 247, 249
Constraints, 159–161, 176, 187, 188, 199, 214,
    215, 218–220, 225
Continuous source of pollutant, 246
Convective acceleration, 44
Conversion factors, 59
Cooling, 14, 15, 200, 203
Coordination model, 220
Correspondence scale, 12
Courant number, 130
Crank–Nicolson scheme, 135, 137

D

Data basis, 231
Data collection, 20, 23, 225, 269, 270
Data retrieval, 21
Data saving (storing), 21

M. Benedini and G. Tsakiris, *Water Quality Modelling for Rivers and Streams*,
Water Science and Technology Library 70, DOI 10.1007/978-94-007-5509-3,
© Springer Science+Business Media Dordrecht 2013

Data screening, 21
Decomposition coefficient, 61
Definitional uncertainty, 237
Density current, 199
Deoxygenation coefficient, 60, 64, 105, 118
Determination coefficient, 228
Deterministic Model, 256
Diffusion number, 131
Dirichlet rule, 132
Discharging population, 58
Dispersion (diffusion), 17, 30–32, 37, 40,
    49–53, 60, 84, 93–96, 104, 107, 110,
    112–114, 117, 128–131, 138, 145,
    146, 150, 161, 177, 187–189, 192,
    193, 200, 201, 209, 210, 237, 238,
    246, 247, 249, 251, 266–268
Dispersion coefficient, 31, 50–53, 145, 146,
    188, 189, 192, 209, 247, 249, 251
Dispersion constant, 51
Dissolved oxygen (DO), 13, 39, 40, 64, 65,
    69–75, 81, 87, 105, 106, 116–123
Divergence theorem, 181
DO. See Dissolved oxygen (DO)
Driver, pressure, state, impact response
    (DPSIR), 7
Dual analysis, 218
Dynamic programming, 219

E
Elementary volume, 28, 29, 36–39, 205
Equivalent population, 58–60, 117
Error, 126, 127, 131, 145, 224, 225, 234, 235,
    238, 251, 253–256, 259, 261, 262
EU directives, 4–5
European Water Framework Directive,
    3, 4, 270
Eutrophication, 8, 78, 87
Explicit numerical schemes, 130, 141–147

F
Falls, 16, 64, 65, 69, 83, 84, 118, 235,
    238, 255
Fickian model, 267, 268
Fick law, 17, 30–32, 37, 112, 266
First order kinetic, 38, 82
Fractals, 268
Fundamental differential equation, 38, 91,
    97, 130, 136, 138, 142, 176, 179,
    185–187, 265
Fuzzy number, 266
Fuzzy set, 266

G
Genetic algorithm, 53, 270
Geographical information system
    (GIS), 270
Growth rate, 72, 81

H
Heat budget, 88
Homogeneous fluid, 32
Hydraulic model, 12, 24, 46, 47, 189,
    219, 241
Hydrodynamic aspects, 41–43
Hydrographic basin, 22, 23
Hyperbolic part, 130, 145
Hyporheic exchange, 267

I
Input-output model, 122
Isoconcentration, 186, 193
Isotropic fluid, 32

J
Joules, 202
Jumps, 64, 115, 250, 267, 268

K
Known terms vector, 119, 120, 159

L
Laminar propagation coefficient, 204
Laws of similitude, 12
Limiting factor, 72
Linear programming, 214–219
Local acceleration, 44
Local injection (source), 32
Local subtraction (sink), 32
Longitudinal dispersion, 31, 50, 51, 53
    coefficient, 51, 53

M
MAE. See Mean absolute error (MAE)
Manning formula, 44, 51
Manning roughness coefficient, 42–45
Mass conservation, 41
Mean absolute error (MAE), 226–227
Measurement uncertainty, 237
Metals and non-metals, 13

Minimum acceptable flow, 16
Momentum conservation, 41
Monitoring systems, 8, 23, 232
Monte Carlo analysis, 255–261
Multicriteria programming model, 219
Mutation, 27, 65

N
Navier–Stokes equation, 41, 42
Neural network, 53, 270
Newton postulate, 204
Nitrate nitrogen, 78
Nitrite nitrogen, 75, 77–78
Nitrogen cycle, 75–78
Non-oscillatory scheme, 182
Normal distribution, 235, 236
Numerical field, 91

O
Objective function, 20, 214, 215, 218, 268
One-dimensional case, 91, 92, 104, 127, 180,
    181, 192
Optimisation (programming) model,
    20, 214
Organic nitrogen, 75–77, 81, 88
Organic pollutants, 82, 83
Oxygen cycle, 69–71
Oxygen deficit, 71, 105, 118
Oxygen demand, 81, 88
Oxygen sag, 106
Oxygen uptake, 73–75, 81

P
Parabolic part, 130, 145
Parameter, 13, 15, 19, 45, 46, 177, 225, 231,
    237, 241, 245–249, 251, 253–256, 259,
    261, 262, 266, 267
Péclet number, 130
Phosphates, 13
Phosphorus cycle, 78–80
Photosynthesis, 72, 73
Plug flow, 105, 115–117
Plume, 185, 186, 199, 201, 265
Pollutant flux, 30
Pollutant wave, 171–176
Pollution types, 57–58
Population (statistical), 234, 258
Probabilistic (stochastic) model, 19

Q
Quality assurance (QA), 239–240
Quality control (QC), 11, 16, 23, 82, 239–240
Quality indicator, 13, 58, 60, 232–234,
    240, 241
QUAL2 model, 137

R
Radiation coefficient, 204
Radiative process, 88
Radionuclides, 83, 84
Random walk, 267, 268
Range of acceptability, 224
Rating curve, 45, 46
Reaction, 13, 27, 32, 38, 39, 60, 65, 69,
    70, 72, 77, 81, 83, 84, 87, 88, 93–95,
    106, 114, 131, 132, 138, 141,
    232, 246
    coefficient, 38, 77, 84, 87, 88, 93–95, 106,
    138, 246
Reaeration coefficient, 40, 62–64, 105,
    117, 118
Redox couple, 81
Respiration, 72, 73, 76, 81
Response curve, 232
Ritz–Galerkin approach, 160–167, 177
River authority, 23
River basin, 2, 4, 6, 22–24, 268–270
Root mean square error (RMSE), 227, 228

S
Saint–Venant equation, 41
Salts, 13, 14
Saturation, 40, 64, 71, 93–95, 105, 106, 117,
    118, 131, 132, 141, 189, 247
Seasonal coefficients, 15
Second Fick Law, 37
Second order accuracy, 130, 145
Sediment oxygen demand (SOD), 73–74
Self-purification process, 15
Settling coefficient, 61
Shallow water equations (SWE), 41–44
Simulation model, 20, 213, 216, 219, 2202
Sink (subtraction), 32
Slack variable, 215
SOD. *See* Sediment oxygen demand (SOD)
Source (injection), 32, 93–95
Specific heat of water, 202, 208
Standard methods, 16

Standard uncertainty, 237
Statistical inference, 261
Steady-state, 103–107, 116, 123, 124
Streeter and Phelps approach, 8, 106
Student probability, 234
Suspended solid, 13, 14, 57, 240–241
SWE. *See* Shallow water equations (SWE)
System science, 19

T

Temperature, 13, 14, 17, 38, 61, 62, 64, 65,
        71, 72, 79, 81, 87–88, 118, 199–204,
        206–211, 241
    adjustment, 87–88
    dependence, 87–88
Thermal plume, 199, 201
Thermal pollution, 199–211, 241
Total organic carbon (TOC), 13, 233
Transverse dispersion, 50
True arithmetic mean, 235
Truncation error, 126, 145
Turbidity, 13, 14, 73, 241
Two-dimensional case, 30, 185–189

U

Uncertainty, 51, 52, 231–242, 250–256, 259,
        261, 262, 266, 267, 269, 270
    error, 238–239, 253, 255

V

Validation, 171, 225, 239, 253, 256, 259, 261
Variables, 11, 17, 19, 20, 44, 72, 73, 92, 98,
        100, 103, 104, 116, 119, 120, 124–126,
        130–132, 142, 150, 151, 181, 194, 209,
        214–218, 234, 235, 237, 238, 249, 251,
        253, 256, 266, 267, 270
Variational form, 195
Verification, 223–228
Virus, 13, 14, 57, 84, 240

W

Washing, 15
Water resources management, 1–8, 11, 21–24,
        213, 219, 220, 268, 270, 271
Water stratification, 87
Weirs, 64–66

Printed by Printforce, the Netherlands